CALIFORNIA EDITION

ASSESSMENT GUIDE
HARCOURT SCIENCE

Harcourt School Publishers

Orlando • Boston • Dallas • Chicago • San Diego

www.harcourtschool.com

Copyright © by Harcourt, Inc.

All rights reserved. No part of this publication may be reproduced or transmitted in any form or by any means, electronic or mechanical, including photocopy, recording, or any information storage and retrieval system, without permission in writing from the publisher.

Teachers using HARCOURT SCIENCE may photocopy test copying masters in complete pages in sufficient quantities for classroom use only and not for resale.

HARCOURT and the Harcourt Logo are trademarks of Harcourt, Inc.

Printed in the United States of America

ISBN 0-15-317692-X

6 7 8 9 10 11 12 13 14 082 2007 2006 2005

Contents

Overview .... AGv
Assessment Components .... AGvi
Formal Assessment .... AGvii
Test-Taking Tips .... AGviii
Performance Assessment .... AGix
Scoring a Performance Task .... AGxi
Classroom Observation .... AGxii
Observation Checklist .... AGxiv
Using Student Self-Assessment .... AGxv
Self-Assessment—Investigate .... AGxvi
Self-Assessment—Learn About .... AGxvii
Project Evaluation Checklist .... AGxviii
Project Summary Sheet .... AGxix
Portfolio Assessment .... AGxx
Science Experiences Record .... AGxxii
Guide to My Science Portfolio .... AGxxiii
Portfolio Evaluation Checklist .... AGxxiv

UNIT A — Living Things Grow and Change

Chapter 1—Plants Grow and Change AG1
Performance Task AG5
Teacher's Directions AG6
Chapter 2—Animals Grow and Change AG7
Performance Task AG11
Teacher's Directions AG12
Chapter 3—People Grow and Change AG13
Performance Task AG17
Teacher's Directions AG18

UNIT B — Exploring Earth's Surface

Chapter 1—Earth's Resources AG19
Performance Task AG23
Teacher's Directions AG24
Chapter 2—Earth Long Ago AG25
Performance Task AG29
Teacher's Directions AG30

UNIT C

Matter and Motion

Chapter 1—Observing and Measuring Matter AG31
Performance Task AG35
Teacher's Directions AG36
Chapter 2—Changes in Matter AG37
Performance Task AG41
Teacher's Directions AG42
Chapter 3—Forces and Motion AG43
Performance Task AG47
Teacher's Directions AG48
Chapter 4—Sound AG49
Performance Task AG53
Teacher's Directions AG54

Extension Chapters

Chapter 1—The Sun, the Moon, and Stars AG55
Performance Task AG59
Teacher's Directions AG60
Chapter 2—Changes in Habitats AG61
Performance Task AG65
Teacher's Directions AG66

Answer Key AG67–AG77

Overview

In *Harcourt Science,* the Assessment Program, like the instruction, is student-centered. By allowing all learners to show what they know and can do, the program provides you with ongoing information about each student's understanding of science. Equally important, the Assessment Program involves the student in self-evaluation, offering you strategies for helping students evaluate their own growth.

The *Harcourt Science* Assessment Program is based on the Assessment Model in the chart below. The model's framework shows the multidimensional aspect of the program, with five kinds of assessment, supported by both teacher-based and student-based assessment tools.

The teacher-based strand, the left column in the model, involves assessments in which the teacher evaluates a student product as evidence of the student's understanding of chapter content and of his or her ability to think critically about it. The teacher-based strand consists of two components: Formal Assessment and Performance Assessment.

The student-based strand, the right column in the model, involves assessments that invite the student to become a partner in the assessment process. These student-based assessments encourage students to reflect on and evaluate their own efforts. The student-based strand also consists of two components: Student Self-Assessment and Portfolio Assessment.

There is a fifth component in the *Harcourt Science* Assessment Program—Ongoing Assessment, which involves classroom observation and informal evaluation of students' growth in science knowledge and process skills. This essential component is listed in the center of the Assessment Model because it is the "glue" that binds together all the other types of assessment.

HARCOURT SCIENCE
Assessment Model

Formal Assessment
- Chapter Review and Test Preparation, PE
- Chapter Test, AG

Ongoing Assessment
- ✓ Questions, PE
- Lesson Review, PE and WB
- Informal Assessment Strategies
  - Observation, TE
  - Performance, TE
  - Portfolio, TE
- Observation Checklist, AG

Student Self-Assessment
- Self-Assessment — Investigate, AG
- Self-Assessment — Learn About, AG
- Project Summary Sheet, AG

Performance Assessment
- Chapter Review, PE
- Chapter Performance Task, AG
- Project Evaluation Checklist, AG

Portfolio Assessment
- Science Experiences Record, AG
- Guide to My Science Portfolio, AG
- Portfolio Evaluation Checklist, AG

(**Key:** PE=Pupil Edition; TE=Teacher's Edition; AG=Assessment Guide; WB=Workbook)

Assessment Components

Formal Assessment

Research into the learning process has shown the positive effects of periodic review. To help you reinforce and assess mastery of chapter objectives, *Harcourt Science* includes both reviews and tests. You will find the Chapter Review and Test Preparation in the pupil book and the Chapter Test in this **Assessment Guide**. Answers to both assessments, including sample responses to open-ended items, are provided.

Performance Assessment

Science literacy involves not only what students know but also how they think and how they do things. Performance tasks provide evidence of students' ability to use science process skills and critical thinking skills to complete an authentic problem-solving task. A performance task is included in each chapter review. Another follows the Chapter Test in this **Assessment Guide**. Each includes teacher directions and a scoring rubric. Also in this booklet, you will find the Project Evaluation Checklist (p. AGxviii), for evaluating unit projects.

Ongoing Assessment

Opportunities abound for observing and evaluating student growth during regular classroom instruction in science. *Harcourt Science* supports this informal, ongoing assessment in several ways: Within each lesson in the **Pupil Edition** (grades 3–5), there are boldface ✔ questions at the end of sections to help you assess students' immediate recall of information. Then, at the end of each lesson, there is a Lesson Review to help you evaluate how well students grasped the concepts taught. The Lesson Review also includes a multiple-choice "test prep" question. In grades 1 and 2, caption questions and Think About It after every lesson are tools for ongoing assessment. Additional material for reviewing the lesson is provided in the **Workbook**.

The **Teacher's Edition** offers Informal Assessment Strategies. These strategies, which appear at point of use within chapters, give ideas for integrating classroom observation, performance assessment, and portfolio assessment with instruction. Located in this **Assessment Guide** is yet another tool, the Observation Checklist (p. AGxiv), on which you can record noteworthy classroom observations.

Student Self-Assessment

Self-assessment can have significant and positive effects on student achievement. To achieve these effects, students must be challenged to reflect on their work and to monitor, analyze, and control their own learning. Located in this **Assessment Guide** are two checklists designed to do just that. One is Self-Assessment—Investigate (p. AGxvi), which leads students to assess their performance and growth in science skills after completing Investigate in the **Pupil Edition**. The second is Self-Assessment— Learn About (p. AGxvii), a checklist to help the student reflect on instruction in a particular lesson or chapter in *Harcourt Science*. Also in this booklet, following the checklists, you will find the Project Summary Sheet (p. AGxix), on which students describe and evaluate their own science projects.

Portfolio Assessment

In *Harcourt Science*, students may create their own portfolios. The portfolio holds self-selected work samples that the student feels represent gains in his or her understanding of science concepts and use of science processes. The portfolio may also contain a few required or teacher-selected papers. Support materials are included in this **Assessment Guide** (pp. AGxx–AGxxiv) to assist you and your students in developing portfolios and in using them to evaluate growth in science skills.

Formal Assessment

Formal assessment is an essential part of any comprehensive assessment program because it provides an objective measure of student achievement. This traditional form of assessment typically consists of reviews and tests that assess how well students understand, communicate, and apply what they have learned. This is the type of assessment that is typically used in state and local standardized tests in science.

Formal Assessment in *Harcourt Science*

Formal assessment in the *Harcourt Science* program includes the following measures: Chapter Review in **Pupil Edition** grades 1 and 2; Chapter Review and Test Preparation in **Pupil Edition** grades 3–5; and Chapter Assessments in this **Assessment Guide.** The purpose of the review is to assess and reinforce not only chapter concepts and science skills but also students' test-taking skills. The purpose of the Chapter Assessments is, as with other formal assessments, to provide an objective measure of student performance. Answers to chapter reviews, including sample responses to open-ended items, are located in the Teacher's Edition, while answers to chapter tests are located in the Answer Key in this booklet.

Types of Review and Test Items

Students can be overwhelmed by the amount of information on a test and uneasy about how to answer different types of test questions about this information. The Chapter Review and Test Preparation is designed to help familiarize students with the various item formats they may encounter: *multiple-choice items* (with a question stem; sentence fragment; graph, table, map, model, or picture; or using negatives such as *not, least,* and so on), *open-ended items* (which require the student to write a short answer, to record data, or to order items), and *scenarios,* in which the student is asked to respond to several items in either a multiple-choice or open-ended format.

Test-Taking Tips

Harcourt Science offers test-taking tips—aimed at improving student performance on formal assessment—in the Teacher's Edition. The section titled Test Prep—Test-Taking Tips spells out what students can do to analyze and interpret multiple-choice or open-ended types of questions. Each tip suggests a strategy that students can use to help them come up with the correct answer to an item. Included in the section are tips to help students

- focus on the question.
- understand unfamiliar words.
- identify key information.
- analyze and interpret graphs, charts, and tables.
- eliminate incorrect answer choices.
- find the correct answer.
- mark the correct answer.

The tips include the following suggestions:

- Scan the entire test first before answering any questions.
- Read the directions slowly and carefully before you begin a section.
- Begin with the easiest questions or most familiar material.
- Read the question and *all* answer options before selecting an answer.
- Watch out for key words such as *not, least,* and so on.
- Double-check answers to catch and correct errors.
- Erase all mistakes completely and write corrections neatly.

Test Preparation

Students perform better on formal assessments when they are well prepared for the testing situation. Here are some things you can do before a test to help your students do their best work.

- Explain the nature of the test to students.
- Suggest that they review the questions at the end of the chapter.
- Remind students to get a good night's sleep before the test.
- Discuss why they should eat a balanced meal beforehand.
- Encourage students to relax while they take the test.

Performance Assessment

Teachers today have come to realize that the multiple-choice format of traditional tests, while useful and efficient, cannot provide a complete picture of students' growth in science. Standardized tests may show what students know, but they are not designed to show how they *think and do things*—an essential aspect of science literacy. Performance assessment, along with other types of assessments, can supply the missing information and help balance your assessment program.

An important feature of performance assessment is that it involves a hands-on activity to solve a situational problem. An advantage of this type of assessment is that students often find it more enjoyable than the traditional paper-and-pencil test. Another advantage is that it models good instruction: students are assessed as they learn and learn as they are assessed.

Performance Assessment in *Harcourt Science*

The performance task, science project, and other hands-on science activities provide good opportunities for performance assessment. The performance task is particularly useful because it provides insights into the student's ability to apply key science process skills and concepts taught in the chapter.

At grades 3–5, *Harcourt Science* provides performance assessment in the Chapter Review and Test Preparation feature in the pupil book and in the Chapter Test in this **Assessment Guide**. In the review at grades 1 and 2, the performance assessment is the last item of the Chapter Review; in the test, it is a performance task. The Project Evaluation Checklist (p. AGxviii) is a measure you can use to evaluate unit projects.

Administering Performance Tasks

Unlike traditional assessment tools, performance assessment does not provide standardized directions for its administration or impose specific time limits on students, although a time frame is suggested as a guideline. The suggestions that follow may help you define your role in this assessment.

▶ *Be prepared.*

A few days before students begin the task, read the Teacher's Directions and gather the materials needed.

▶ *Be clear.*

Explain the directions for the task; rephrase them as needed. Also, explain how students' performance will be evaluated. Present the rubric you plan to use and explain the performance indicators in language your students understand.

▶ *Be encouraging.*

Your role in administering the assessments should be that of a coach—motivating, guiding, and encouraging students to produce their best work.

▶ *Be supportive.*

You may assist students who need help. The amount of assistance needed will depend on the needs and abilities of individual students.

▶ *Be flexible.*

All students need not proceed through the performance task at the same rate and in the same manner. Allow them adequate time to do their best work.

▶ *Involve students in evaluation.*

Invite students to join you as partners in the evaluation process, particularly in development or modification of the rubric.

Rubrics for Assessing Performance

A well-written rubric can help you score students' work accurately and fairly. Moreover, it gives students a better idea of what qualities their work should exhibit *before* they begin a task.

Each performance task in the program has its own rubric. The rubric lists performance indicators, which are brief statements of what to look for in assessing the skills and understandings that the task addresses. A sample rubric follows.

Scoring Rubric

Performance Indicators

_____ Assembles kite successfully.

_____ Carries out experiment daily.

_____ Records results accurately.

_____ Makes an accurate chart and uses it to report the strength of wind observed each day.

Performance Indicators

| 3 | 2 | 1 | 0 |

Scoring a Performance Task

The scoring system used for program performance tasks is a 4-point scale (3-2-1-0) that is compatible with those used by many state assessment programs. You may wish to modify the rubrics as a 3- or 5-point scale, as your individual needs and circumstances require. To determine a student's score on a performance task, review the indicators checked on the rubric and then select the score that best represents the student's overall performance on the task.

4-Point Scale			
Excellent Achievement	Adequate Achievement	Limited Achievement	Little or No Achievement
3	2	1	0

How to Convert a Rubric Score into a Grade

If, for grading purposes, you want to record a letter or numerical grade rather than a holistic score for the student's performance on a task, you can use the following conversion table:

Holistic Score	Letter Grade	Numerical Grade
3	A	90–100
2	B	80–89
1	C	70–79
0	D–F	69 or below

Developing Your Own Rubric

From time to time, you may want to either develop your own rubric or work together with your students to create one. Research has shown that significantly improved performance can result from student participation in the construction of rubrics.

Developing a rubric for a performance task involves three basic steps: (1) Identify the process skills taught in the chapter that students must perform to complete the task successfully and identify what understanding of content is also required. (2) Determine which skill/understanding is involved in each step. (3) Decide what you will look for to confirm that the student has acquired each skill and understanding you identified.

Classroom Observation

"Kid watching" is a natural part of teaching and an important part of evaluation. The purpose of classroom observation in assessment is to gather and record information that can lead to improved instruction. In this booklet, you will find an Observation Checklist on which you can record noteworthy observations of students' ability to use science process skills.

Using the Observation Checklist

▶ *Identify the skills you will observe.*
Find out which science process skills are introduced and reinforced in the chapter.

▶ *Focus on only a few students at a time.*
You will find this more effective than trying to observe the entire class at once.

▶ *Look for a pattern.*
It is important to observe the student's strengths and weaknesses over a period of time to determine whether a pattern exists.

▶ *Plan how and when to record observations.*
Decide whether to
- record observations immediately on the checklist as you move about the room or
- make jottings or mental notes of observations and record them later.

▶ *Don't agonize over the ratings.*
Students who stand out as particularly strong will clearly merit a rating of *3* ("Outstanding"). Others may clearly earn a rating of *1* ("Needs Improvement"). This doesn't mean, however, that a *2* ("Satisfactory") is automatically the appropriate rating for the rest of the class. For example, you may not have had sufficient opportunity to observe a student demonstrate certain skills. The checklist cells for these skills should remain blank under the student's name until you have observed him or her perform the skills.

▶ *Review your checklist periodically and ask yourself questions such as:*

> *What are the student's strongest/weakest attributes?*
>
> *In what ways has the student shown growth?*
>
> *In what areas does the class as a whole show strength/weakness?*
>
> *What kinds of activities would encourage growth?*
>
> *Do I need to allot more time to classroom observation?*

▶ *Use the data you collect.*

Refer to your classroom observation checklists when you plan lessons, form groups, assign grades, and confer with students and family members.

Observation Checklist

Date _____

AG xiv Assessment Guide

Rating Scale

- ☒ 3 Outstanding
- ☒ 1 Needs Improvement
- ☒ 2 Satisfactory
- ☐ Not Enough Opportunity to Observe

Names of Students

Science Process Skills

- Observe
- Compare
- Classify/Order
- Gather, Record, Display, or Interpret Data
- Use Numbers
- Communicate
- Plan and Conduct Simple Investigations
- Measure
- Predict
- Infer
- Draw Conclusions
- Use Time/Space Relationships
- Hypothesize
- Formulate or Use Models
- Identify and Control Variables
- Experiment

Grade 2

Using Student Self-Assessment

Researchers have evidence that self-evaluation and the reflection it involves can have positive effects on students' learning. To achieve these effects, students must be challenged to reflect on their work and to monitor, analyze, and control their own learning—beginning in the earliest grades.

Frequent opportunities for students to evaluate their performance build the skills and confidence they need for effective self-assessment. A trusting relationship between the student and the teacher is also essential. Students must be assured that honest responses can have only a positive effect on the teacher's view of them, and that they will not be used to determine grades.

Student Self-Assessment in *Harcourt Science*

The assessment program offers three self-assessment measures, which are located in this booklet. The first one is Self-Assessment—Investigate: a form that invites students to reflect on how they felt about, and what they learned from, Investigate, a hands-on investigation at the beginning of each lesson. The second measure is Self-Assessment—Learn About: a form that leads students to reflect on and evaluate what they learned from reading and instruction in Learn About at either the lesson or chapter level. The third is the Project Summary Sheet—a form to help students describe and evaluate their unit projects.

Using Self-Assessment Forms

▶ *Explain the directions.*
  Discuss the forms and how to complete them.

▶ *Encourage honest responses.*
  Be sure to tell students that there are no "right" responses to the items.

▶ *Model the process.*
  One way to foster candid responses is to model the process yourself, including at least one response that is not positive. Discuss reasons for your responses.

▶ *Be open to variations in students' responses.*
  Negative responses should not be viewed as indicating weaknesses. Rather they confirm that you did a good job of communicating the importance of honesty in self-assessment.

▶ *Discuss responses with students.*
  You may wish to clarify students' responses in conferences with them and in family conferences. Invite both students and family members to help you plan activities for school and home that will motivate and support their growth in science.

Name _____

Date _____

Self-Assessment—Investigate

How Did I Do?

Investigate was about

_____

_____

How did you do? Circle the word that tells what you think. If you are not sure, circle the **?**.

1. I followed the directions.   Yes   ?   No
2. I worked well with others.   Yes   ?   No
3. I was careful with materials.   Yes   ?   No
4. I completed the investigation.   Yes   ?   No
5. The science skill that I learned was

[   ]

6. I found out

_____

AG xvi Assessment Guide   Grade 2

Name _____

Lesson _____

Self-Assessment—Learn About

Think Back!

How did you do? Circle the word that tells what you think. If you are not sure, circle the **?**.

1. I could read the lesson. **Yes ? No**
2. I used the pictures to help me read. **Yes ? No**
3. I answered the questions by the pictures. **Yes ? No**
4. I asked questions when I did not understand something. **Yes ? No**
5. I understood most of the ideas. **Yes ? No**
6. I could answer most of the questions in Think About It. **Yes ? No**

This is something I learned.

_____

_____

I learned these new words.

_____

_____

Grade 2

Name: _____
Date: _____

Project Evaluation Checklist

Project Evaluation

Aspects of Science Literacy	Evidence of Growth
1. **Understands science concepts** (Animals, Plants; Earth's Land, Air, Water, Space; Matter, Motion, Energy)	_____ _____ _____
2. **Uses science process skills** (observes, compares, classifies, gathers/interprets data, communicates, measures, experiments, infers, predicts, draws conclusions)	_____ _____ _____ _____
3. **Thinks critically** (analyzes, synthesizes, evaluates, applies ideas effectively, solves problems)	_____ _____ _____
4. **Displays traits/attitudes of a scientist** (is curious, questioning, persistent, precise, creative, enthusiastic; uses science materials carefully; is concerned for environment)	_____ _____ _____ _____

Summary Evaluation/Teacher Comments: _____
_____
_____
_____
_____

AG xviii Assessment Guide   Grade 2

Name _____
Date _____

Project Summary Sheet

You can tell about your science project by completing the following sentences.

My Unit Project

1. My project was about _____
   _____
   _____
   _____.

2. I worked on this project with _____
   _____
   _____
   _____.

3. I gathered information from these sources: ___
   _____
   _____
   _____.

4. The most important thing I learned from doing this project is _____
   _____
   _____
   _____.

5. I think I did a (an) _____ job on my project because
   _____
   _____
   _____.

6. I'd also like to tell you _____
   _____
   _____
   _____
   _____
   _____.

Grade 2

Assessment Guide AG xix

Portfolio Assessment

A portfolio is a showcase for student work, a place where many types of assignments, projects, reports, and writings can be collected. The work samples in the collection provide "snapshots" of the student's efforts over time, and taken together they reveal the student's growth, attitudes, and understanding better than any other type of assessment. However, portfolios are not ends in themselves. Their value comes from creating them, discussing them, and using them to improve learning.

The purpose of using portfolios in science is threefold:

▶ *To give the student a voice in the assessment process.*

▶ *To foster reflection, self-monitoring, and self-evaluation.*

▶ *To provide a comprehensive picture of a student's progress.*

Portfolio Assessment in *Harcourt Science*

In *Harcourt Science,* students create portfolio collections of their work. The collection may include a few required papers, such as the Chapter Test, Chapter Performance Task, and Project Evaluation.

From time to time, consider including other measures (Science Experiences Record, Project Summary Sheet, and Student Self-Assessment Checklists). The Science Experiences Record, for example, can reveal insights about student interests, ideas, and out-of-school experiences (museum visits, nature walks, outside readings, and so on) that otherwise you might not know about. Materials to help you and your students build portfolios and use them for evaluation are included in the pages that follow.

Using Portfolio Assessment

▶ *Explain the portfolio and its use.*
Describe how people in many fields use portfolios to present samples of their work when they are applying for a job. Tell students that they can create their own portfolio to show what they have learned, what skills they have acquired, and how they think they are doing in science.

▶ *Decide what standard pieces should be included.*
Engage students in identifying a few standard, or "required," work samples that each student should include in his or her portfolio, and discuss reasons for including them. The student's recording sheet for the Chapter Performance Task, for example, might be a standard sample in the portfolios because it shows students' ability to use science process skills and critical thinking skills to solve a problem. Together with your class, decide on the required work samples that everyone's portfolio will include.

▶ *Discuss student-selected work samples.*
Point out that the best work to select is not necessarily the longest or the neatest. Rather, it is work the student believes will best demonstrate his or her growth in science understanding and skills.

▶ *Establish a basic plan.*
Decide about how many work samples will be included in the portfolio and when they should be selected. Ask students to list on Guide to My Science Portfolio (p. AG xxiii) each sample they select and to explain why they selected it.

▶ *Tell students how you will evaluate their portfolios.*
Use a blank Portfolio Evaluation sheet to explain how you will evaluate the contents of a portfolio.

▶ *Use the portfolio.*
Use the portfolio as a handy reference tool in determining students' science grades and in holding conferences with them and family members. You may wish to send the portfolio home for family members to review.

Name _____ Date _____

Science Experiences Record

Date	What I Did	What I Thought or Learned

GUIDE TO MY
Science Portfolio

Name _____ Date _____

What Is in My Portfolio	Why I Chose It
1.	
2.	
3.	
4.	
5.	
6.	
7.	

I organized my Science Portfolio this way because _____
_____
_____
_____
_____

Grade 2 **Assessment Guide AG xxiii**

Student's Name _____

Date _____

Portfolio Evaluation Checklist

Portfolio Evaluation

Aspects of Science Literacy	Evidence of Growth
1. **Understands science concepts** (Animals, Plants; Earth's Land, Air, Water, Space; Matter Motion, Energy)	_____ _____ _____
2. **Uses science process skills** (observes, compares, classifies, gathers/interprets data, communicates, measures, experiments, infers, predicts, draws conclusions)	_____ _____ _____ _____
3. **Thinks critically** (analyzes, synthesizes, evaluates, applies ideas effectively, solves problems)	_____ _____ _____
4. **Displays traits/attitudes of a scientist** (is curious, questioning, persistent, precise, creative, enthusiastic; uses science materials carefully; is concerned for environment)	_____ _____ _____ _____

Summary of Portfolio Assessment

For This Review			Since Last Review		
Excellent	Good	Fair	Improving	About the Same	Not as Good

AG xxiv Assessment Guide

Grade 2

Name _____
Date _____

People Grow and Change

Part 1   Vocabulary

Draw a line from each label to the correct body part.

1. lungs •

2. heart •

3. permanent teeth •

4. stomach •

5. skeleton •

6. muscles •

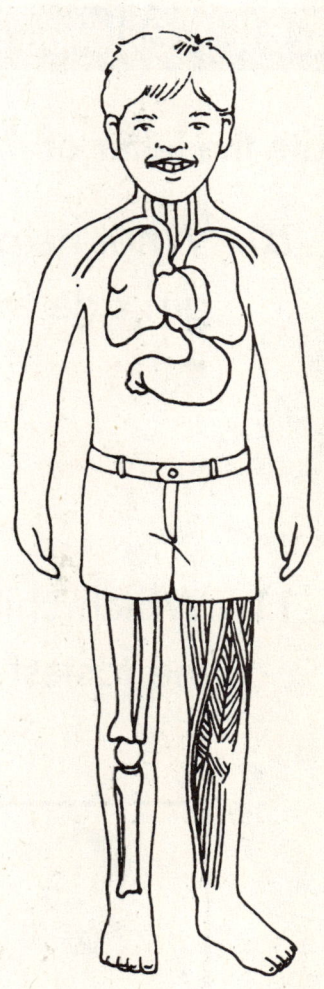

Unit A • Chapter 3   Assessment Guide   AG 13

Name _____

Write the letter of the best choice.

    A digest   **B** heart rate   **C** saliva

7. Your ___ gets faster when you exercise.

8. To get energy, your body must ___ food.

9. The liquid in your mouth that breaks down food is ___.

Part II   Science Concepts and Understanding

Write the letter of the best choice.

___ 10. What helps this girl ride fast?

   A helmet
   B light
   C saliva
   D heart

___ 11. Which child would have the fastest heart rate?

   F   **G**   **H**   **J**

Name _____

12. Write an *H* below the healthful food.

A

C

___   ___

B

D

___   ___

Write the name of the body part below the picture of something it works with.

muscles   stomach   heart

13. 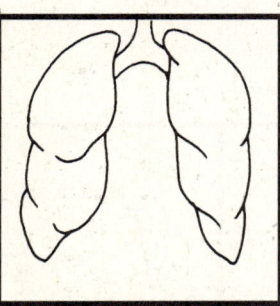   14. 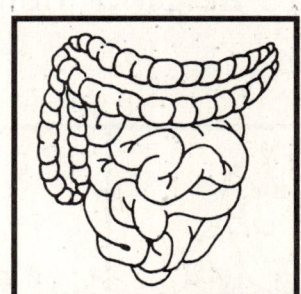   15.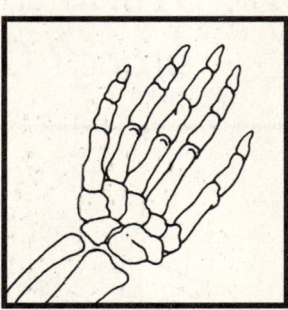

___   ___   ___

Unit A • Chapter 3   (page 3 of 4)   Assessment Guide AG 15

Name _____

| Part III   Process Skills Application |

Process skills: *compare, observe*

16. Tell how the girl has changed from Picture A to Picture B.

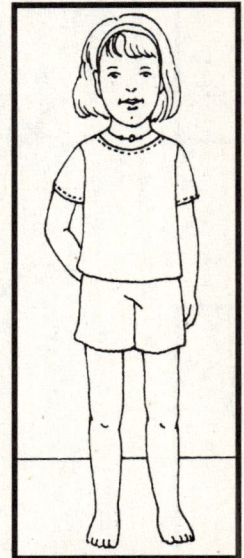

Picture A   **Picture B**

_____

_____

17. Which children are helping their bodies grow strong? Circle them.

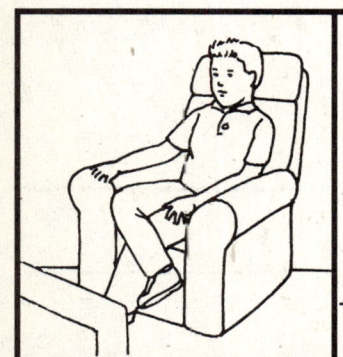   

AG 16 Assessment Guide   Unit A • Chapter 3

Name _____ Date _____

Growing Healthy and Strong

PERFORMANCE TASK

Materials

index card with body part name on it

crayons

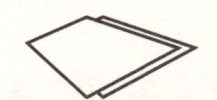
drawing paper

1. Get an index card with the name of a body part on it.

2. Think about what this body part does for you.

3. Think of a healthful activity you like to do that makes this body part grow strong.

4. Draw a picture that shows you doing this activity. Write a sentence that tells why it is healthful for the body part. Label the parts of your picture.

5. Explain your picture to a group of classmates.

Unit A • Chapter 3   Assessment Guide   AG 17

PERFORMANCE TASK

Teacher's Directions

Growing Healthy and Strong

Materials — Performance Task sheets, index cards, crayons, drawing paper

Time — 15–20 minutes

Suggested Grouping — individuals or pairs

Science Processes — observe, compare, communicate

Preparation Hints — Prepare enough index cards for each child to have one. Write *bones, muscles, heart,* or *lungs* on each card.

Introduce the Task — Have four volunteers stand in front of the class, each holding a different body-part card. Ask why each of these body parts is important, and discuss the things we do to keep these parts healthy. Distribute the Performance Task sheets, and have volunteers read the instructions.

Promote Discussion — Have children join small groups and explain why the activity they have shown is healthful for the body part on their card.

Scoring Rubric

Performance Indicators

____ Draws a picture of a healthful activity (for example, playing baseball or eating broccoli).

____ Draws a picture of an activity that is related to the body part on his or her card.

____ Labels his or her drawing and tells why the activity is healthful for the body part.

____ Uses words such as *skeleton, heart rate,* and *heartbeat,* and/or refers specifically to how the body part works.

Observations and Rubric Score

| 3 | 2 | 1 | 0 |

Unit A • Chapter 3

Name _____

Date _____

Chapter Assessment

Earth's Resources

Part 1   Vocabulary

Draw a line from each word to the picture it labels.

1. boulder •

2. sand •

3. soil •

4. mineral •

5. transportation •

Write the letter of the answer that best completes the sentence.

6. A ___ is a hard, nonliving thing that comes from Earth.

7. Something that people can use is called a ___.

8. A ___ is something found in nature that people can use to meet their needs.

A natural resource

B rock

C resource

Unit B • Chapter 1   (page 1 of 4)   Assessment Guide AG 19

Name _____

Part II Science Concepts and Understanding

9. Put an **X** beside the natural resources.

 ____    ____

 ____    ____

10. Write the name of one kind of transportation that uses water.

_____

11. Write the name of one kind of transportation that uses air.

_____

Name _____

Write the letter of the best choice.

___ 12. What do people make from [trees] ?
   A minerals
   B cars
   C furniture
   D animals

___ 13. What are bricks made from?
   F clouds   G soil   H trees   J copper

14. Complete the chart with the names of natural resources.

WAYS WE USE NATURAL RESOURCES

Natural Resource	Product
_____	(coins)
_____	(fruit bowl)
_____	(kite)

Unit B • Chapter 1   (page 3 of 4)   Assessment Guide AG 21

Name _____

Part III Process Skills Application

Process skills: make a model, collect data

15. Draw a model of something that can be made from rock. Then write a label that describes the model.

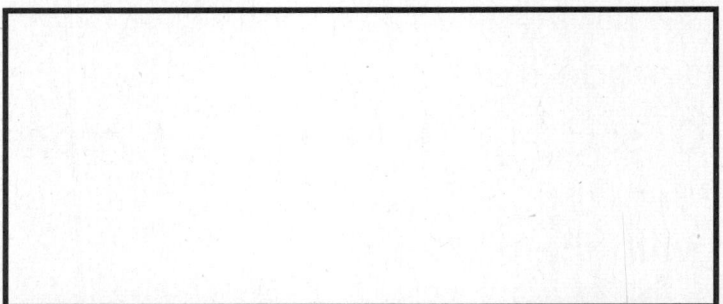
_____

16. Complete the table below the picture.

WORKING IN THE GARDEN	
Number of people using water	
Number of people using soil	
Number of people using rock	

AG 22 Assessment Guide   (page 4 of 4)   Unit B • Chapter 1

Name _____ Date _____

What Is Thrown Away?

PERFORMANCE TASK

Materials

pencil   construction paper   markers

1. Write *What Is Thrown Away?* on a sheet of paper. Put it near the wastebasket. Ask classmates to list what they throw away. Stop after 10 items.

2. Classify the items into three groups. Count the items in a tally table. Make a graph of the information.

3. Use your graph to explain to the class what resources are being thrown away.

What Is Thrown Away?
1. ____
2. ____
3. ____
4. ____
5. ____
6. ____
7. ____
8. ____

What Is Thrown Away?

Group	Number of Items
paper	1 2 3 4 5 6
plastic	
other	

Tally Table

paper	
plastic	
other	

Unit B • Chapter 1   Assessment Guide   AG 23

PERFORMANCE TASK

Teacher's Directions

What Is Thrown Away?

Materials — Performance Task sheets, paper, pencil

Time — 20–30 minutes

Suggested Grouping — individuals or small groups

Science Processes — gather, record, display, and interpret data; use numbers

Preparation Hints — Write the words *soil, water, trees,* and *minerals* on index cards and display on chalk tray.

Introduce the Task — Promote a discussion of natural resources by asking children to brainstorm ways people use soil, water, trees, and minerals. Make sure children know that these are natural resources. Remind them of the need to use these resources carefully. Distribute materials. Ask a volunteer to read the first direction, and relate it to the picture of the list. Ask other volunteers to read the second and third directions, and relate them to the pictures. Ensure that all children understand how to proceed. Establish a way for children to take turns displaying their lists near the wastebasket.

Promote Discussion — After children finish, have them compare their graphs with other children's. Promote a discussion about what kind of natural resources were being thrown away in the wastebasket.

Scoring Rubric

Performance Indicators

_____ Elicits help from other children to fill in the *What Is Thrown Away?* list.

_____ Accurately tallies items in a tally table.

_____ Prepares graph of tallied information.

_____ Uses the graph to explain the types and amounts of resources that are being thrown away.

Observations and Rubric Score

3   2   1   0

Name _____
Date _____

Earth Long Ago

Part I Vocabulary

Draw a line from each word to the picture it names.

1. paleontologist •

2. dinosaur •

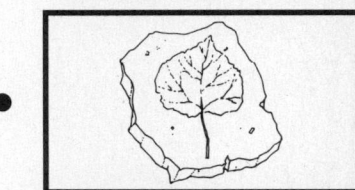

3. fossil •

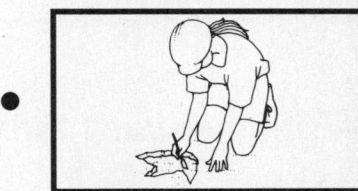

Write the letter of the best answer.

___ 4. Scientists try to rebuild, or ___, dinosaur skeletons from fossils.

    A grow   **B** reconstruct   **C** name   **D** picture

___ 5. We can't see live dinosaurs today because they are ___.

    F extinct   **G** eating   **H** too big   **J** eggs

Name _____

___ 6. ___ means "three-horned face."

    A Brontosaurus   **C** Triceratops
    B Tyrannosaurus rex   **D** Stegosaurus

Part II   Science Concepts and Understanding

7. Circle the one that is **NOT** a fossil.

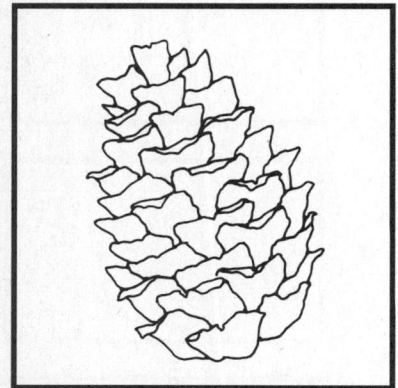 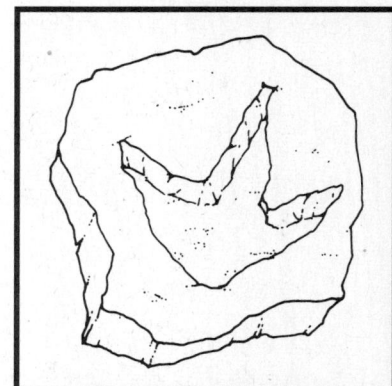 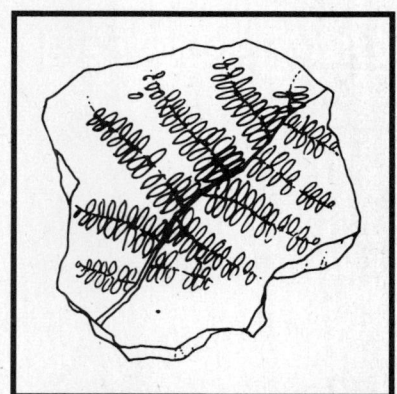

Write the word that matches each picture.

Paleontologists find fossils in tar.
Where else do they find fossils?

    rock   trees   amber

8.
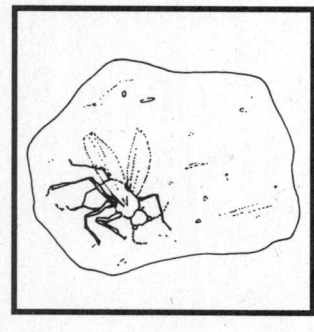
in _____

9.
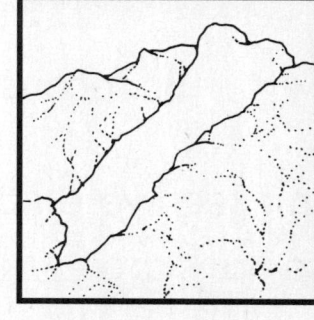
in _____

Name _____

10. Number the pictures to show the correct order.

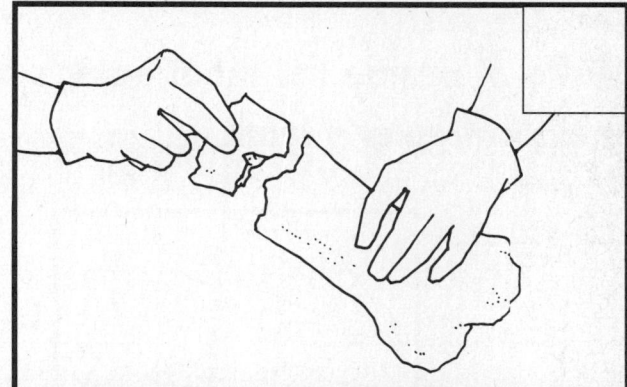
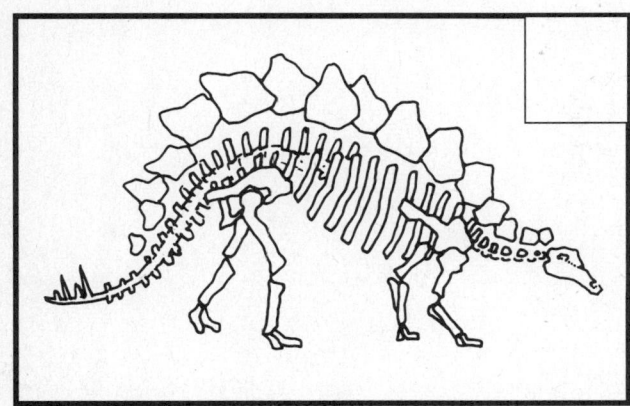

Circle the letter of the best answer.

11. What does the word *dinosaur* mean?

   A terrible chicken
   B lazy lizard
   C terrible lizard
   D lucky lizard

12. Where did some dinosaurs lay their eggs?

   F in the water
   G in rocks
   H in trees
   J in nests

Unit B • Chapter 2   (page 3 of 4)   Assessment Guide   AG 27

Name _____

Part III   Process Skills Application

Process skills: make a model, infer

Write *M* beside the meat eater's teeth and *P* beside the plant eater's teeth.

13. ___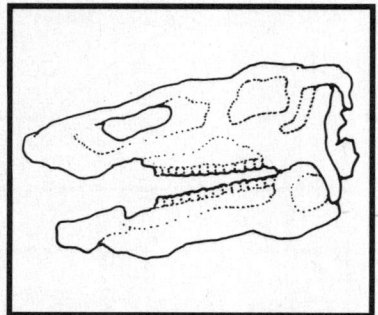

14. ___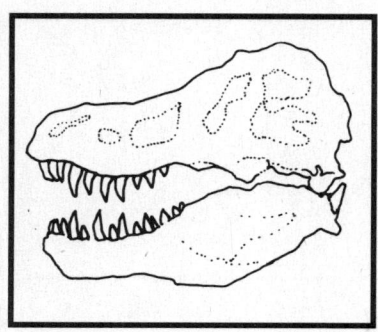

15. Draw a fossil of a part of an extinct plant.

16. Draw a picture to show what it looked like as a living thing.

Name _____ Date _____

Searching for Treasure

PERFORMANCE TASK

Materials

chocolate chip cookies   toothpick   small paintbrush   paper

Paleontologists learn about the past by digging up and studying fossils. Think about how a paleontologist gets fossils out of rocks without breaking them.

You are a paleontologist. Your job is to get the chips out of your cookie without breaking them.

1. BE CAREFUL. Toothpicks are sharp. Use your tools to remove the chips.

2. Count the number of whole chips you get out of your cookie.

3. Record your number of chips on a class chart.

4. Make a graph to show how many whole chips were found in each cookie.

Unit B • Chapter 2   Assessment Guide   AG 29

PERFORMANCE TASK

Teacher's Directions

Searching for Treasure

Materials Performance Task sheets, chocolate chip cookies, toothpicks, small paintbrushes, paper

Time 20 minutes

Suggested Grouping individuals and small groups

Science Processes model, infer, communicate

Preparation Hints Use packaged cookies for this activity. They are harder and more rocklike than homemade cookies. If the cookies are too hard to work with, try blueberry muffins. Have children wash their hands. Have extra cookies for them to eat later. Make a class chart for children to use to record the number of whole chips they get out.

Introduce the Task Ask children to think about what it would be like to dig for fossils. Let volunteers draw on the board a tool they think would help them and explain how they would use it. Read and discuss the directions with children. Remind them that toothpicks are sharp. Tell them not to eat the cookies as they work; they can have one later.

Promote Discussion When children have finished, have them compare graphs. Elicit ideas on how their task was like or different from that of a scientist on a fossil hunt.

Scoring Rubric

Performance Indicators

_____ Uses toothpick to pick away big pieces and paintbrush to brush away small crumbs.

_____ Records number of chips on class chart.

_____ Draws accurate graph to show number of whole chips found in each cookie.

_____ Shows qualities of a paleontologist such as persistence, patience, and the ability to stick to the task.

Observations and Rubric Score

3   2   1   0

Name _____
Date _____

Observing and Measuring Matter

Part 1   Vocabulary

Draw a line from each word to the picture it goes with.

1. centimeter •   •

2. milliliter •   •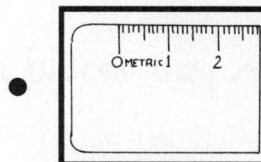

3. mass •   •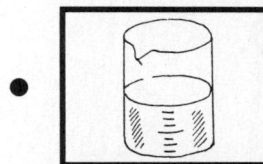

4. solid •   •

5. liquid •   •

6. gas •   •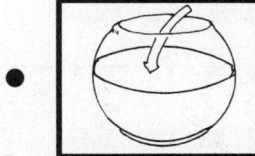

Unit C • Chapter 1   (page 1 of 4)   Assessment Guide   AG 31

Name _____

Write the letter of the word that completes the sentence.

A solid   **B** property   **C** matter

7. Solids, liquids, and gases are different kinds of ___.

8. A quality that something has is called a ___.

Part II   Science Concepts and Understanding

Write **C** below the things you would measure using centimeters.

Write **M** below the things you would measure using milliliters.

9.   10.   11.

___   ___   ___

12. What kind of matter is air? How do you know?

_____

_____

Name _____

Write the letter of the correct answer.

___ 13. What kind of matter can be measured in milliliters?

A gas
B texture
C solid
D liquid

___ 14. What kind of matter has a shape of its own?

F gas
G solid
H liquid
J property

___ 15. What kind of matter can you measure using two balloons and a rod?

A gas
B solid
C liquid
D bubble

16. How can you change the shape of a solid?

_____

_____

Name _____

Part III Process Skills Application

Process skills: classify, measure

17. Write **S** for solid, **L** for liquid, or **G** for gas under each picture below.

_____

_____

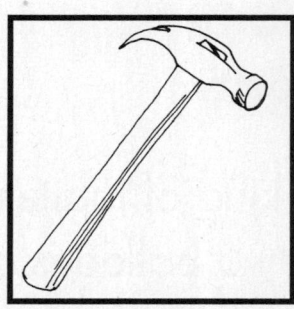

_____

_____

18. Write an **X** under the balance holding more mass.

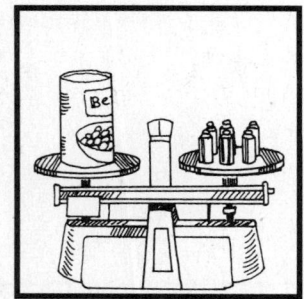

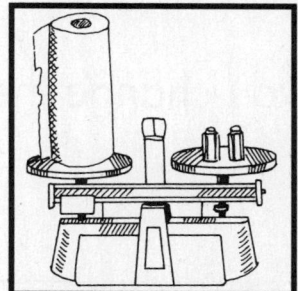

_____

_____

Name _____ Date _____

Solid or Liquid?

Materials

 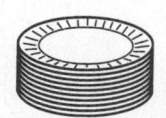 

play slime   paper plate   paper cup

1. Take a spoonful of slime, and try four tests. After each, decide whether the slime has the properties of a solid or a liquid. Record your observations in the chart below.

Slime Tests

	Test	Solid or Liquid?	How I Know
A.	Roll into a ball.		
B.	Hold in your hand.		
C.	Pour into a cup.		
D.	Set on a paper plate.		

2. Think about how the slime changes. Does the slime have the properties of a solid or a liquid? Explain how you know.

PERFORMANCE TASK

Teacher's Directions

Solid or Liquid?

Materials — Performance Task sheets, play slime, paper plates, paper cups

Time — 30 minutes

Suggested Grouping — individuals

Science Processes — conduct an investigation, communicate, infer

Preparation Hints — To make the slime, mix 2 parts cornstarch with 1 part colored water. Be sure to color the water first. Mixture should be runny but thick enough to push or roll into a ball. Put a spoonful of slime on a paper plate for each child.

Introduce the Task — Ask children to name some liquids and solids and tell how they know. (Children may say they can pour a liquid or hold a solid in their hands.) Distribute Performance Task sheets, and ask children to read the directions. Answer any questions children raise. Then give them the materials for the slime tests.

Promote Discussion — When children finish, have them share their findings in small groups and discuss whether slime is liquid, solid, or both. Have each group report its conclusion to the class.

Scoring Rubric

Performance Indicators

_____ Conducts the tests and records results on chart.

_____ Uses appropriate descriptions of a solid and a liquid when explaining how he or she knows what the slime is. May say that slime is a liquid because it flows to take the shape of a container or that it is a solid because it holds its shape when rolled into a ball.

_____ Records observations after each test.

_____ Infers that slime has the characteristics of both a solid and a liquid.

Observations and Rubric Score

| 3 | 2 | 1 | 0 |

Name _____
Date _____

Changes in Matter

Part 1   Vocabulary

Write words from the box to describe the pictures.

| irreversible change   mixture   reversible change |

1.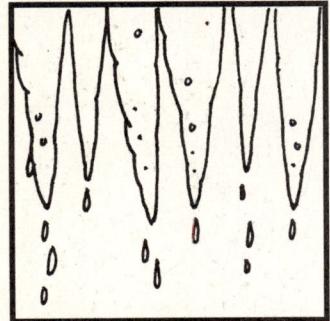

_____

2.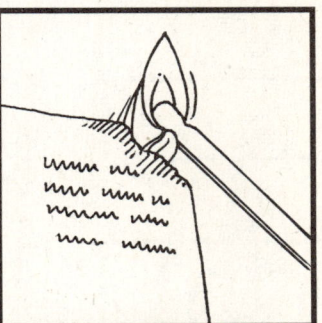

_____

3.

_____

Unit C • Chapter 2   Assessment Guide   AG 37

Name _____

Part II Science Concepts and Understanding

There is a form of water in each picture. Write the letter of the form under the picture.

S = solid   L = liquid   G = gas

4.

_____

5.

_____

6.

_____

7.

_____

8. Circle all the things that change the shape of matter but do not change its mass.

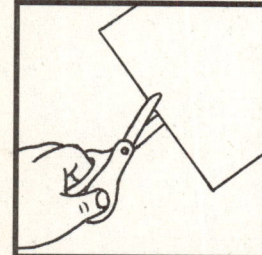 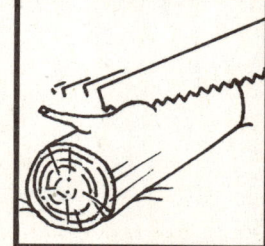 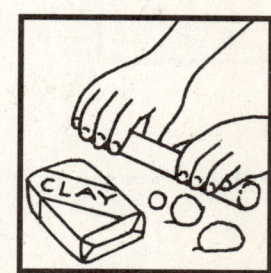

AG 38 Assessment Guide   Unit C • Chapter 2

Name _____

Write the letter of the best answer.

___ 9. At which temperature will water stay a solid?
   A 0°C   C 23°C
   B 10°C   D 32°C

___ 10. What kind of change happens in a forest fire?
   F mixture   H reversible
   G logging   J irreversible

Write the answer to the question.

11. What can you add or take away from water to make it change its form?

_____

12. How can you make an irreversible change to an egg?

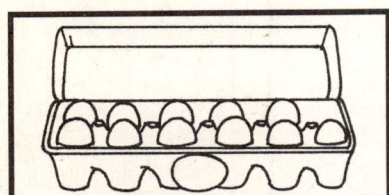

_____

Name _____

Part III   Process Skills Application

Process Skills: observe/classify, plan an investigation, predict

13. Circle the things you might use to change a liquid into a gas.

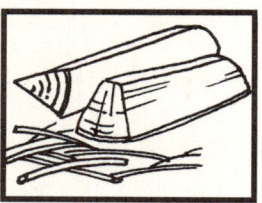   

 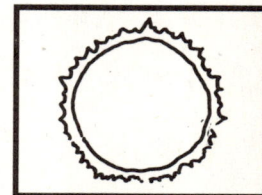 

14. List the steps you would take to change the liquid into a gas.

_____

_____

15. Look at each picture. Predict which kind of change will happen. Write *reversible* or *irreversible* under the picture.

AG 40 Assessment Guide   Unit C • Chapter 2

Name _____ Date _____

How Does Water Change?

Materials

writing paper   drawing paper   crayons

1. Look at the pictures. How could you use some of these things to show how water changes? Plan an investigation.

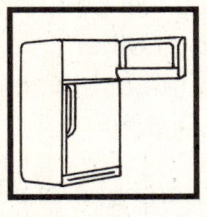      

2. Circle the things you would use.

3. Write a plan for your investigation. Draw pictures to show what you would do.

4. Predict what would happen if you conducted your experiment. Share your plan with others.

Unit C • Chapter 2   Assessment Guide   AG 41

PERFORMANCE TASK

Teacher's Directions

How Does Water Change?

Materials — Performance Task sheets, writing paper, drawing paper, crayons, cards with terms printed on them: *solid, liquid, gas, water vapor, rain, snow, ice, heat*

Time — 20–30 minutes

Suggested Grouping — individuals and small groups

Science Processes — observe, plan an investigation, predict

Preparation Hints — Prepare cards for the eight terms.

Introduce the Task — Place the cards where children can see them. Ask three volunteers to hold up the cards for the three states of matter (*liquid, solid, gas*). Then ask volunteers to select cards for what we call water when it is a gas (*water vapor*), a solid (*ice, snow*), and a liquid (*rain*). Display the card for *heat*. Ask children how heat can change water (take it away from rain and you get ice; add it to ice and you get water). Tell children they will be thinking about how water can be changed from one form of matter to another. Distribute Performance Task sheets, and have children read the directions silently. Ask volunteers to explain how they will follow the directions.

Promote Discussion — Organize children in groups with others who planned investigations for the same change in form (for example, from a liquid to a solid). Ask children to compare the investigations they planned.

Scoring Rubric

Performance Indicators

_____ Circles all the necessary items on Performance Task sheet.

_____ Plans an investigation to change water from one state of matter to another.

_____ Tells what might happen if he or she were to carry out the investigation.

_____ Shares and compares plans with other children.

Observations and Rubric Score

3   2   1   0

Name _____

Date _____

Forces and Motion

Part 1   Vocabulary

Write the correct word to complete the sentence.

| force | gravity | location | motion | wind |

The girl with the wagon is pulling it, or using
1. _____ to move it. She is
changing its 2. _____ by moving
it toward the trees. The force of moving air, or
3. _____, keeps the kite up in the
sky. The force that pulls one end of the seesaw
down is called 4. _____. When
the seesaw is moving up and down, we say it is
in 5. _____ .

Unit C • Chapter 3   (page 1 of 4)   Assessment Guide   AG 43

Name _____

Part II   Science Concepts and Understanding

Write the word that describes the force used in each picture.

6. _____

7. _____

8. _____

Name _____

9. Write an *X* on the picture that shows something a magnet cannot do.

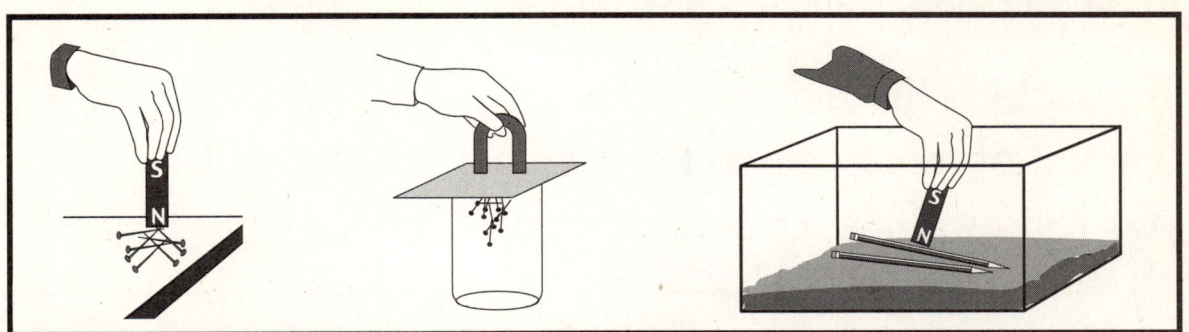

Write the letter of the correct answer.

___ 10. What makes it harder to pull a wagon on grass than on a sidewalk?
    A magnetism   C wind
    B friction   D motion

___ 11. What can you use to measure how long it takes to run a race?
    F rubber band   H stopwatch
    G meterstick   J location

___ 12. When you push or pull something, what might you change?
    A its rough surface   C its gravity
    B its location   D its color

___ 13. What is the force that keeps us on the ground?
    F location   H gravity
    G wind   J grass

Unit C • Chapter 3   (page 3 of 4)   Assessment Guide   AG 45

Name _____

Part III Process Skills Application

Process skills: observe, use numbers

The scale in the pictures below measures the amount of force. When the hammer hits the scale, you can read the amount of force.

A

B

_____   _____

14. Write an **X** below the picture that shows the greater force.

15. Use information from the pictures to complete the table.

Person	Amount of Force
A	
B	

Name _____ Date _____

If There Were No Gravity...

PERFORMANCE TASK

Materials

masking tape   pencils   meterstick or ruler   paper   crayons or markers

1. Mark a line on the floor with tape. Put your toes on the line and jump as far in front of it as you can. Measure each jump with a meterstick or ruler.

2. Make a table to record each child's jump.

3. Think about the force that moves each child up in the air and the force that brings that child down. Then complete the table below.

Force	Motion
Legs push against floor	
	Child comes back to floor

4. Draw a picture of yourself jumping in place where there is no gravity. Describe your picture.

Unit C • Chapter 3   Assessment Guide   AG 47

PERFORMANCE TASK

Teacher's Directions

If There Were No Gravity...

Materials Performance Task sheets, masking tape, pencils, metersticks or rulers, paper, crayons or markers

Time 20–30 minutes

Suggested Grouping individuals or small groups

Science Processes observe, measure, use numbers, predict

Preparation Hints You may wish to take children outside to jump. Use chalk to mark a line on the playground.

Introduce the Task Ask children what it would be like to travel in space. Discuss the absence of gravity in space (without gravity, everything floats; nothing is held down). Distribute the Performance Task sheets. Ask volunteers to read the directions aloud. Make sure that children understand the task. Distribute the remaining materials.

Promote Discussion When children have finished, have them meet in small groups. Ask them to compare their drawings and paragraphs. Then ask children to decide as a group how they would measure a broad jump on a spacecraft. Ask each group to report to the class the method its members agreed upon for measuring the broad jump.

Scoring Rubric

Performance Indicators

_____ Measures and records jumps.

_____ Completes the table on force and motion.

_____ Draws and interprets a picture of jumping in a place with no gravity.

_____ Writes an explanation of what it would be like to jump in a place that has no gravity.

Observations and Rubric Score

| 3 | 2 | 1 | 0 |

Name _____
Date _____

Sound

Part 1 Vocabulary

Draw a line from the word to the picture it goes with.

1. loudness •   •

2. vibrate •   •

3. sonar •   •

4. music •   •

Write the word that best completes the sentence.

| pitch   sound |

5. Energy that you hear is _____.

6. A whistle has a high _____.

Name _____

Part II Science Concepts and Understanding

7. Why are the plates vibrating?

_____

_____

8. Why are the worker's ears covered?

_____

_____

Write the letter of the best choice.

___ **9.** What vibrates when we speak?
  A vocal cords
  B eardrums
  C inner ears
  D throat

___ **10.** What is the pitch of a mouse's squeak?
  F low
  G high
  H soft
  J loud

___ **11.** What vibrates when we hear?
  A throat
  B vocal cords
  C eardrums
  D ear lobes

___ **12.** Which animal does **NOT** sing?
  F canary
  G butterfly
  H humpback whale
  J human being

Name _____

Is the sound traveling through a liquid, a solid, or a gas?

13. _____

14. _____

15. What does an audiologist test?
    _____

16. What special way of using sound helps dolphins find things?
    _____

17. Write 1, 2, 3, 4 to show the order in which things happen.

    ___ The sound waves move through the air.

    ___ A person strums the strings of a guitar.

    ___ Someone hears guitar music.

    ___ The strings of the guitar vibrate.

Name _____

Part III Process Skills Application

Process skills: compare, observe, plan an investigation

18. You can tap a solid surface with two parts of a marker or a pencil and compare the sounds. Complete the chart. The two rows for the marker have been done.

SOUND ON A SOLID SURFACE

Tap	Part	Louder	Softer	Higher	Lower
marker	cap		X	X	
	bottom	X			X
pencil	eraser				
	point				

19. How could you find out whether a large drum or a small drum has a lower pitch? Plan an investigation to answer this question.

_____
_____
_____
_____

Name _____ Date _____

Tapping a Tune

PERFORMANCE TASK

Materials

four glasses   spoon   water

1. Fill the glasses with different amounts of water.

2. Tap the side of each glass gently with the spoon and listen to the sound it makes.

3. Investigate what happens when you put more water in a glass. Does the pitch get higher or lower?

4. Investigate what happens to the pitch when you pour some water back into the pitcher.

5. Add or take away water to change the sounds until you can tap out a tune.

6. Tap your tune for the class.

PERFORMANCE TASK

Teacher's Directions

Tapping a Tune

Materials — Performance Task sheets, four glasses, water, pitcher, spoon

Time — 30 minutes

Suggested Grouping — pairs or small groups

Science Processes — observe, compare, conduct an investigation

Preparation Hints — Set up a work site and pour different amounts of water into two of the glasses in order to introduce the task.

Introduce the Task — Ask a volunteer to tap each glass gently with a spoon. Ask children to decide which sound has the higher pitch. Invite volunteers to hum or sing the first part of a tune the pitch brings to mind. Distribute Performance Task sheets, and ask children to read the directions silently. Then ask volunteers to tell what they will be doing. Explain that children are not to handle the glasses. They are to change the pitch only by adding water from the pitcher or taking water out a spoonful at a time, placing that water back into the pitcher. When students finish, have each group tap out its tune.

Promote Discussion — Ask children to share what they learned about changing the pitch of a glass of water. They should explain how to make the pitch higher and lower by adding and taking away water.

Scoring Rubric
Performance Indicators

_____ Fills glasses with different amounts of water.

_____ Listens to sounds of glasses when tapped, deciding which are higher or lower.

_____ Investigates changing the sound by adding or removing water.

_____ Taps out a four-pitch tune on the glasses.

Observations and Rubric Score

3   2   1   0

Name _____
Date _____

The Sun, the Moon, and Stars

Part 1   Vocabulary

Write the correct word to complete each sentence.

| constellation   sun   moon   craters   energy   rotation |

1. The closest star to Earth is the _____.

2. The sun gives off two kinds of _____.

3. The spinning of Earth is called its _____.

4. The object that looks the largest to us in the night sky is the _____.

5. The moon has large holes called _____.

6. A group of stars that forms a star picture is a _____.

Circle the word that completes the sentence.

7. The sun's light and heat are called ___.

   orbit   solar energy   craters

Extension • Chapter 1   (page 1 of 4)   Assessment Guide   AG 55

Name _____

8. Earth moves in a path, or an ___, around the sun.

astronaut   **energy**   **orbit**

9. Different ___ have different kinds of weather.

constellations   **seasons**   **craters**

10. The light from the moon, or ___, comes from the sun.

moonlight   **dust**   **rotation**

Part II   Science Concepts and Understanding

For Questions 11–12, use the picture of Earth and the sun.

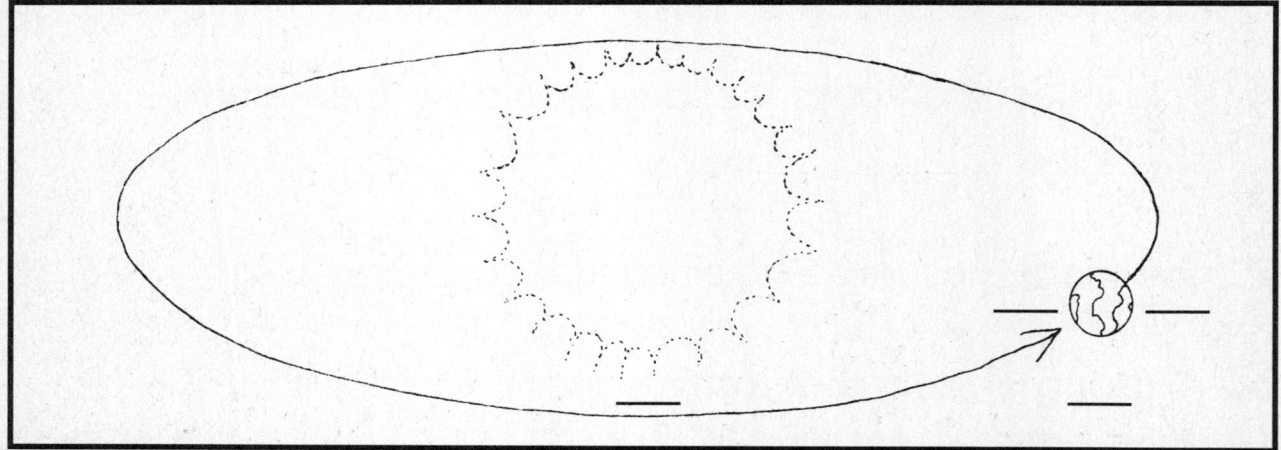

11. Write **E** under Earth and **S** under the sun.

12. On one side of Earth, it is day. On the other side, it is night. Write **D** for day and **N** for night.

Name _____

13. What is the sun made of?

_____

Write the letter of the correct answer.

___ **14.** How long does it take Earth to orbit the sun?
     A ten minutes   **C** one month
     B one hour   **D** one year

___ **15.** What season is it when the sun hits our part of Earth most directly?
     F spring   **H** fall
     G summer   **J** winter

16. Draw a line from each word to the thing it describes.

rocks •

astronaut •

crater •

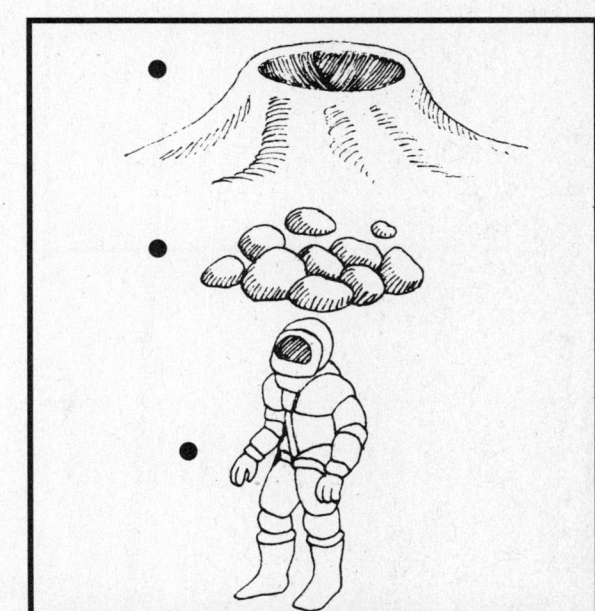

Name _____

Part III   Process Skills Application

Process skills: time/space relationships, compare

17. Write an **X** under the new moon.

  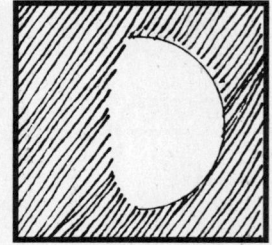 

___   ___   ___   ___

18. Circle the constellation we call the Big Dipper.

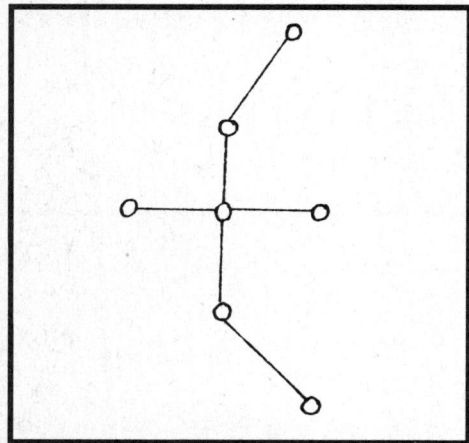 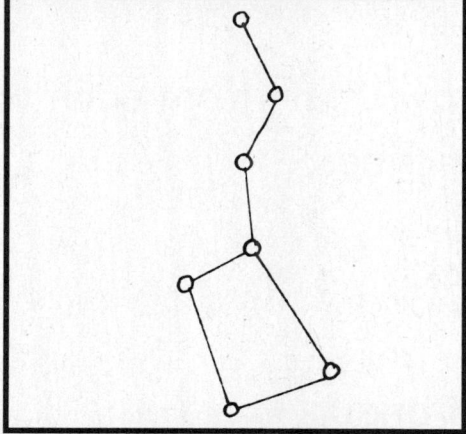

Name _____ Date _____

Night and Day

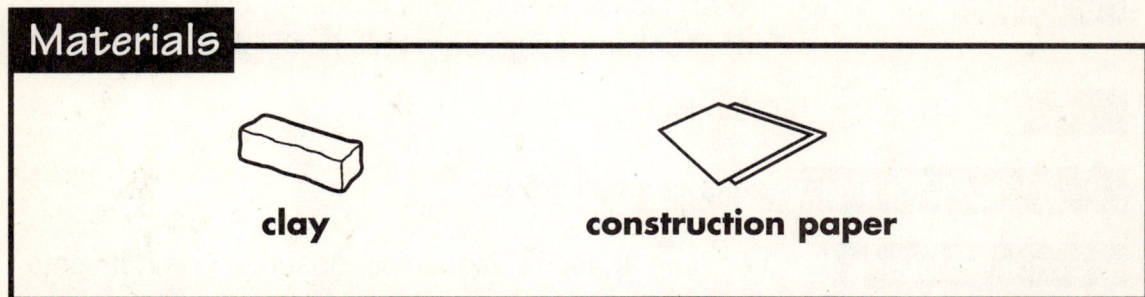

Materials: clay, construction paper

Work with a partner.

1. Make clay models of the sun and Earth.

2. Talk about how Earth's rotation gives us night and day.

3. One partner holds the sun. The other partner holds Earth.

4. The person holding Earth moves around the sun. Spin Earth to show how it rotates while it orbits the sun.

5. Draw a picture of the sun and Earth. Label the parts of your picture. Write **Night** by the dark side of Earth and **Day** by the lighted side.

Extension • Chapter 1

PERFORMANCE TASK

Teacher's Directions

Night and Day

Materials Performance Task sheets, modeling clay, construction paper, foam ball, flashlight

Time 20 minutes

Suggested Grouping pairs or small groups

Science Processes use time/space relationships, observe, communicate

Preparation Hints When you divide the clay, be sure each pair or group has a piece big enough to make a model of Earth and the sun.

Introduce the Task Introduce the activity by having one volunteer hold the foam ball and another the flashlight. Aim the flashlight (the sun) at the ball, and ask which part of the ball (Earth) is having day and which is having night. Have the volunteer rotate the ball, and repeat the question. Ask children to speculate about the parts of Earth experiencing night and day. Distribute the Performance Task sheets. Have volunteers read the directions aloud, and have children explain in their own words what they will be doing. Then distribute the materials.

Promote Discussion When children finish, help them recall how long it takes Earth to rotate one time (one day) and how long it takes Earth to orbit the sun (one year).

Scoring Rubric
Performance Indicators

_____ Creates reasonable models of the sun and Earth, making the sun much larger than Earth.

_____ Works cooperatively to simulate Earth's rotation and orbit.

_____ Rotates Earth while moving it in an orbit around the sun.

_____ Draws a picture of the sun and Earth and labels night and day correctly.

Observations and Rubric Score

3   2   1   0

AG 60 Assessment Guide Extension • Chapter 1

Name _____
Date _____

Changes in Habitats

Part 1 Vocabulary

Draw a line from each word to the picture it goes with.

1. drought   2. litter   3. recycle

Write the letter of the word that best completes the sentence.

4. Waste that hurts land, water, or air is called ___.

   A drought   C environment
   B pollution   D recycle

5. You can ___ tires to make a playground.

   F reuse   G change   H burn   J litter

Extension • Chapter 2   (page 1 of 4)   Assessment Guide AG 61

Name _____

Part II   Science Concepts and Understanding

6. Write what changed this habitat.

_____

Write a word that describes a kind of pollution.
Then draw a picture of it.

7. 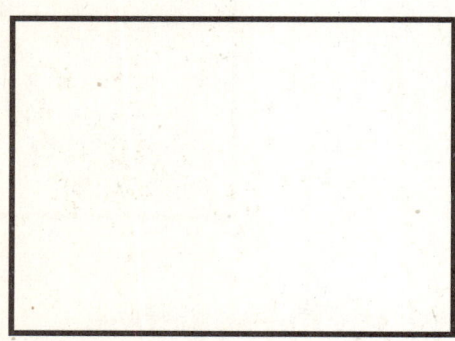   8.

_____

Circle the letter of the best choice.

9. What makes air pollution?

   A rain   C plants
   B cars   D animals

Name _____

10. What is litter?

    F grass   **H** plants
    G water   **J** trash

11. Circle the person who is helping the environment.

12. Draw someone doing something else to help the environment. Write about what the person is doing.

_____

_____

Name _____

Part III Process Skills Application

Process skills: draw conclusions, compare

13. Write about something you see here that is hurting the environment.

_____

_____

14. Put an **X** beside the things that you can recycle or reuse.

A

C

B

D

AG 64 Assessment Guide   (page 4 of 4)   Extension • Chapter 2

Name _____ Date _____

How Pollution Hurts

Materials

animal picture card

habitat picture card

crayons

construction paper

1. Get one picture card of an animal and one of its habitat.

2. Think about what could happen to the animal if its habitat became polluted.

3. Make a sign. Draw a picture to show what might happen.

4. Write a saying about the problem.

5. Explain your picture and saying to a partner. Then explain your partner's sign to the class.

Extension • Chapter 2

PERFORMANCE TASK

Teacher's Directions

How Pollution Hurts

Materials   Performance Task sheets, animal picture cards, habitat picture cards, construction paper, crayons

Time   20–30 minutes

Suggested Grouping   individuals or pairs

Science Processes   draw conclusions, communicate

Preparation Hints   Put together sets of habitat picture cards and animal picture cards, ensuring that each set has an animal that lives in the habitat.

Introduce the Task   Have a volunteer name a habitat. Then have the class name animals that live in that habitat. Ask volunteers to name the parts of a habitat that can be damaged by pollution (water, air, land). Write their responses on the board. Then distribute the Performance Task sheets. Have volunteers read the directions aloud. Distribute the remaining materials.

Promote Discussion   When children finish, ask them to share their sign with a partner. The partner will identify the type of pollution the sign is about. Each child then either confirms the partner's conclusion or explains what the sign means. Then have children display and explain their partner's sign. If possible, post the signs in the classroom.

Scoring Rubric

Performance Indicators

____ Draws appropriate environmental hazard.
____ Shows the effects of the pollution on an animal.
____ Writes an appropriate message for his or her sign.
____ Explains a classmate's sign clearly to the class.

Observations and Rubric Score

| 3 | 2 | 1 | 0 |

Unit A • Chapter 1 • Plants Grow and Change

Plants Grow and Change

Part I Vocabulary 4 points each

Write the word that best completes each sentence.

germinates   living   nonliving   nutrients

1. Things that grow and change are
   living.

2. The air and water that animals need are
   nonliving.

3. The minerals that plants get from the soil are
   nutrients.

4. Growing begins when a seed
   germinates.

Circle the letter of the word or words that tell what changed the way these plants grew.

9. A root (B) insects C flower D seed coat

10. F rain G soil (H) light J leaves

Write *cactus*, *oak tree*, or *pine tree* to name each plant.

11. pine tree
12. cactus
13. oak tree

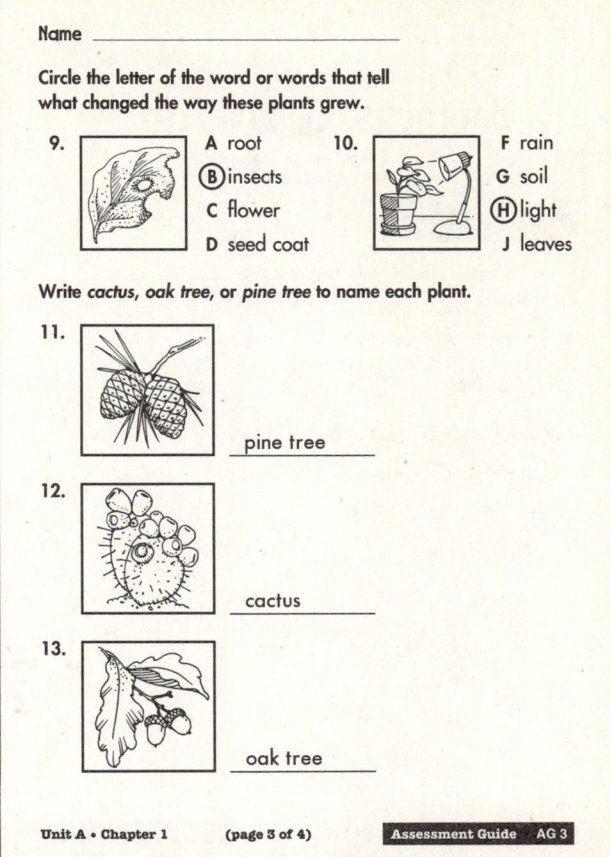

Draw a line from each label to the picture it names.

5. seedling
6. seed coat

Part II Science Concepts and Understanding 8 points each

Complete the chart.
Put an X in each box that shows what the living thing needs.

What Living Things Need

Living Thing	Water	Light	Air
7.	X	—	X
8.	X	X	X

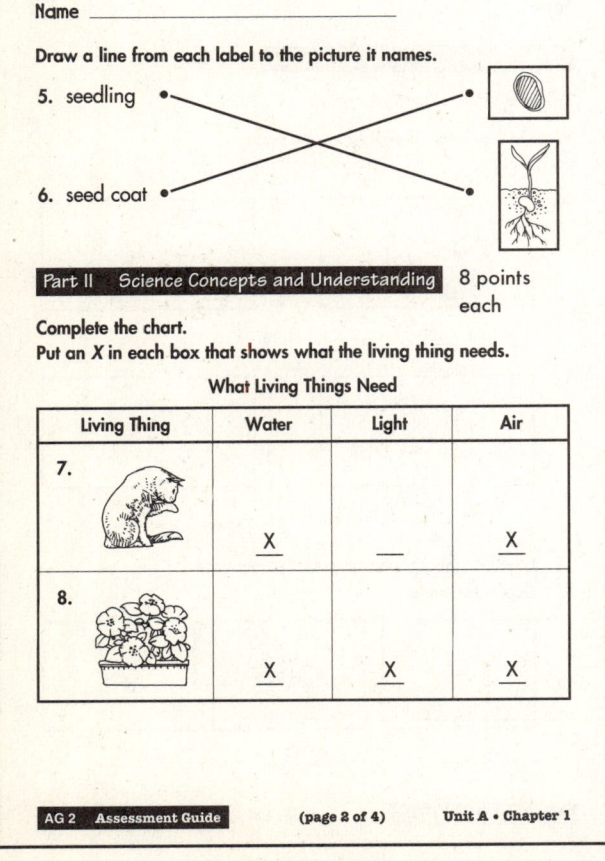

Part III Process Skills Application 10 points each

Process skills: observe, classify

14. Write *l* for **living** or *n* for **nonliving** by each thing. The first one is done for you.

   l   n
   n   l

15. Put an X by the seed that is germinating.

   X (bottom box)

Answer Key

Unit A • Chapter 2 • Animals Grow and Change

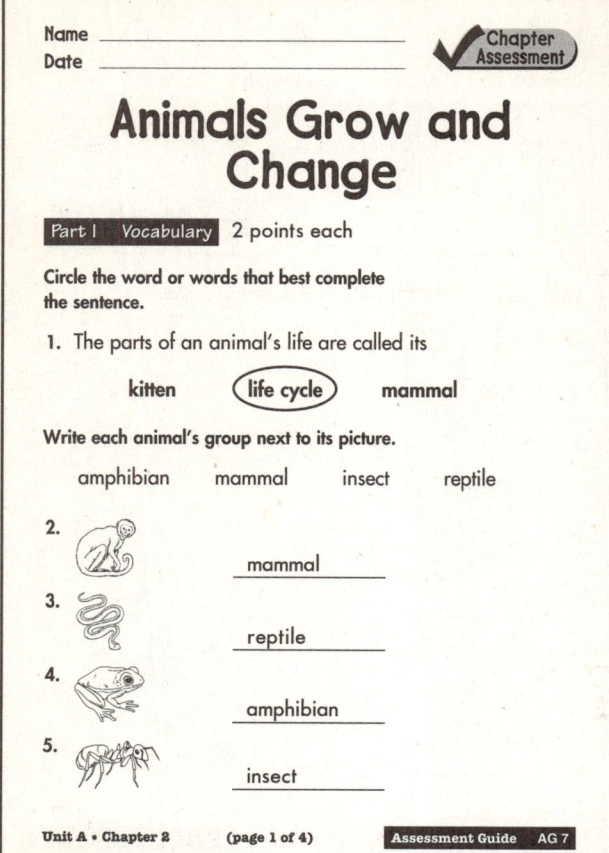

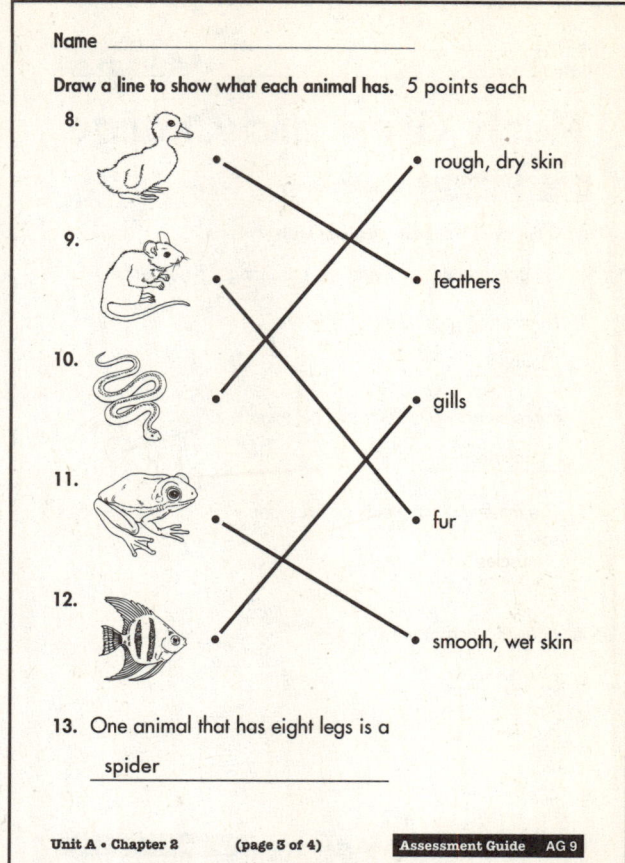

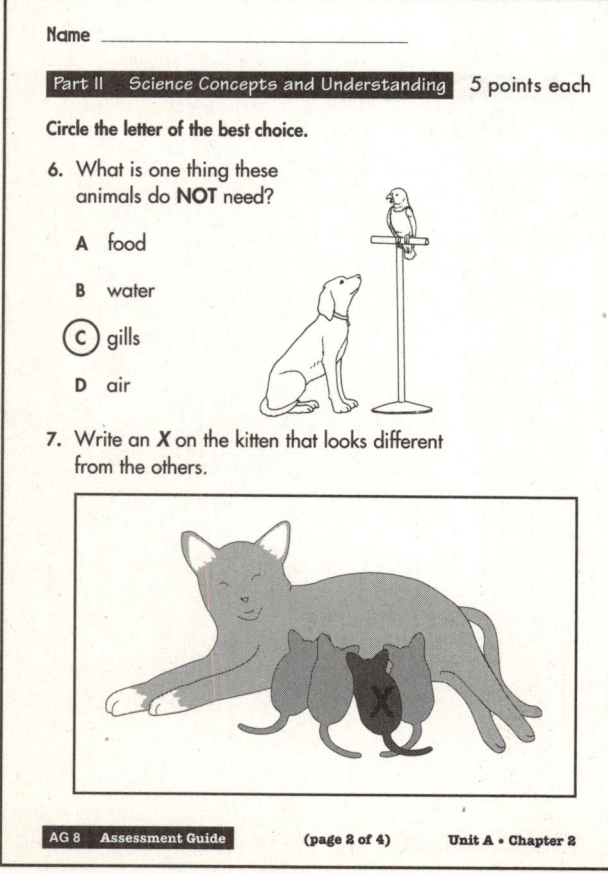

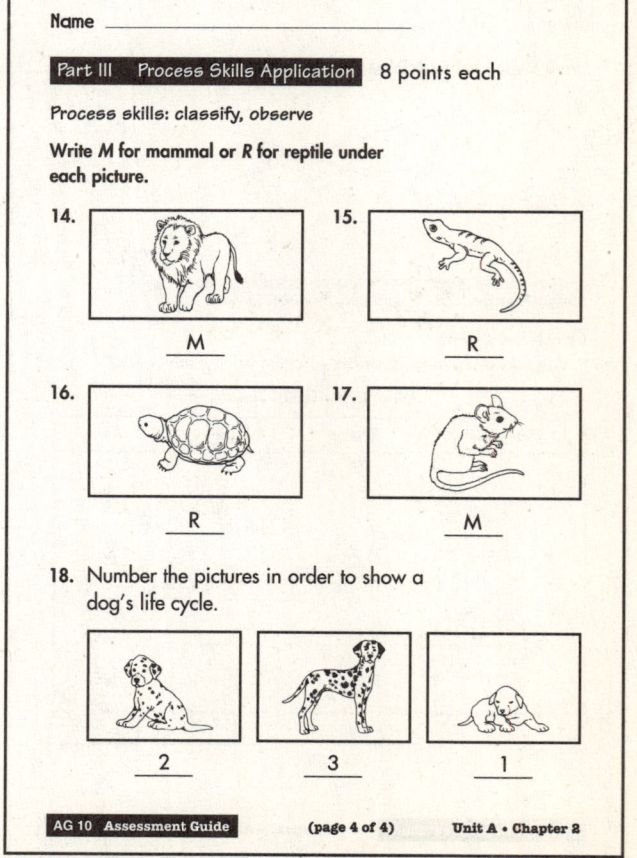

AG 68 Assessment Guide

Answer Key

Unit A • Chapter 3 • People Grow and Change

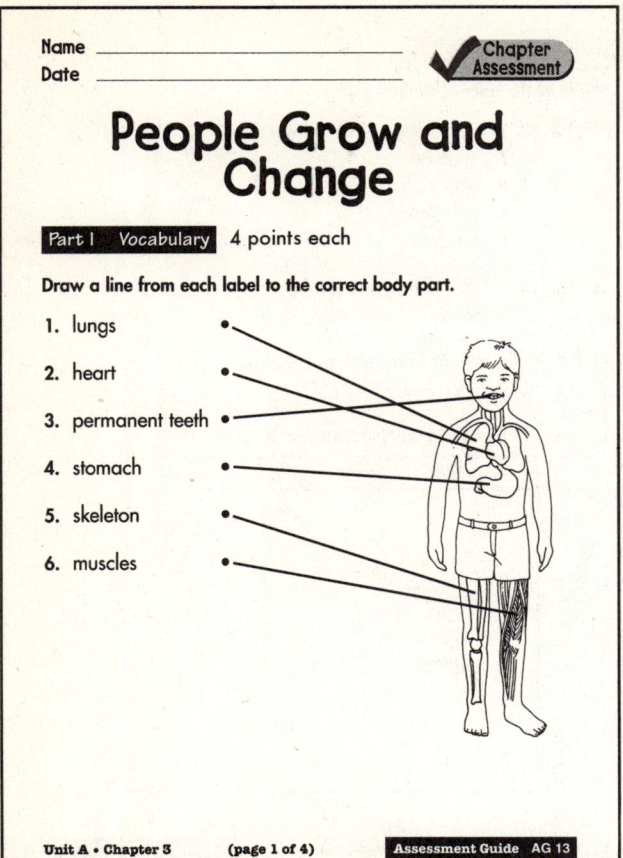

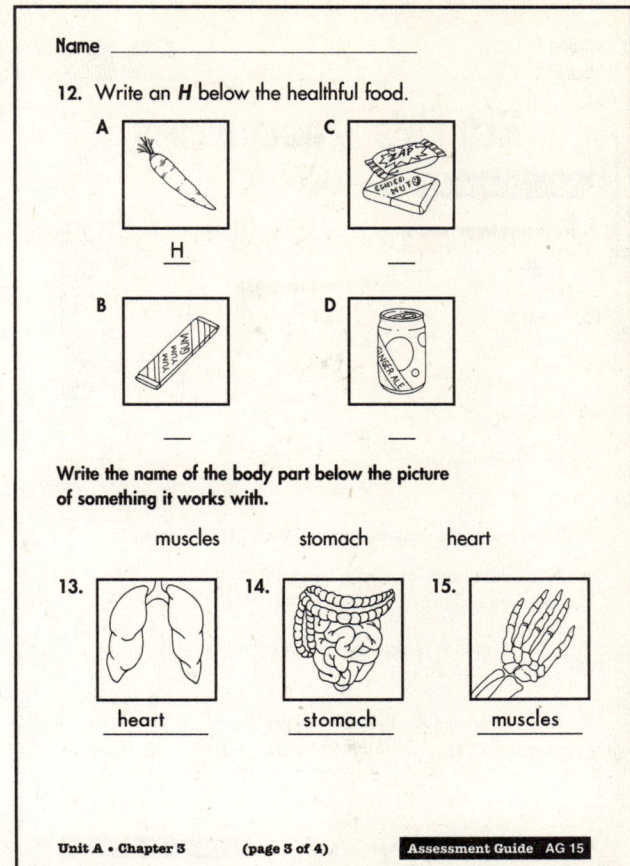

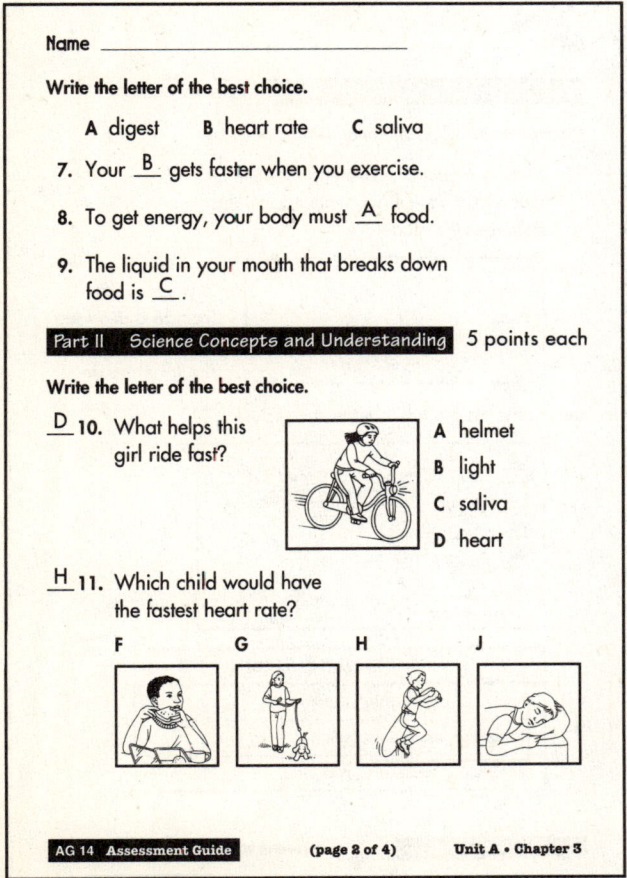

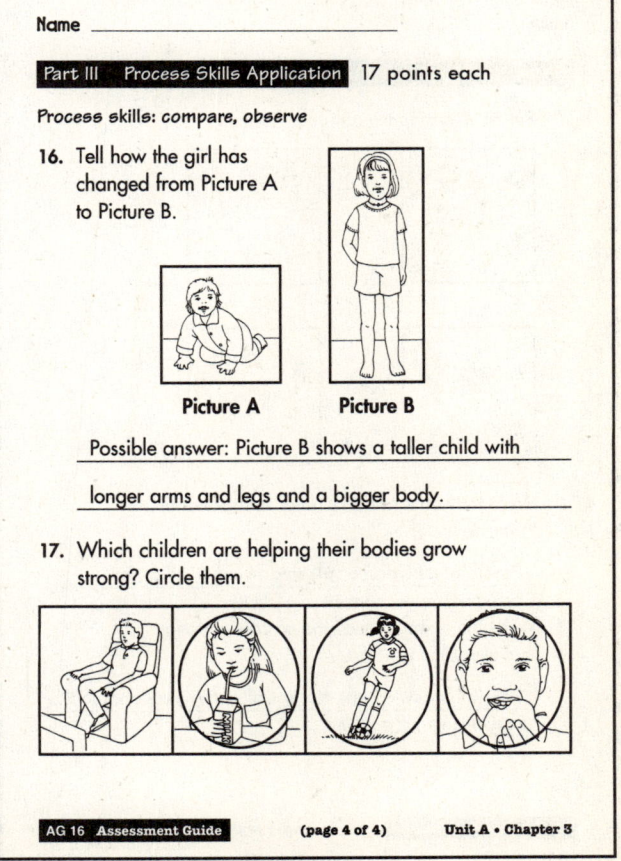

Answer Key

Assessment Guide AG 69

Unit B • Chapter 1 • Earth's Resources

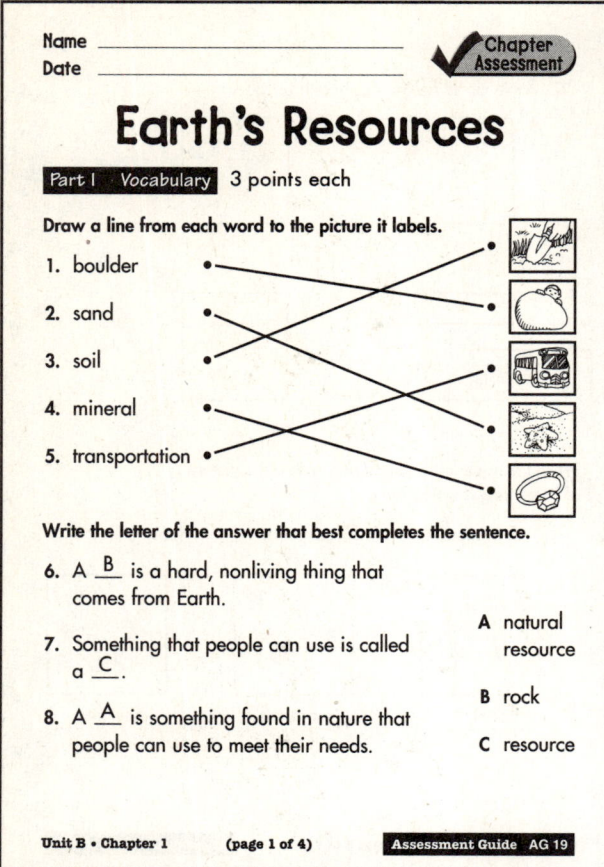

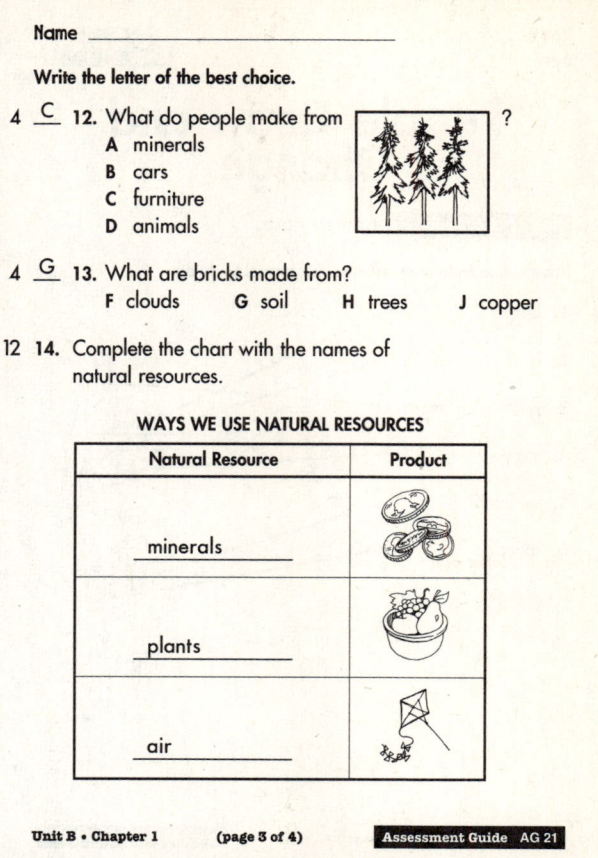

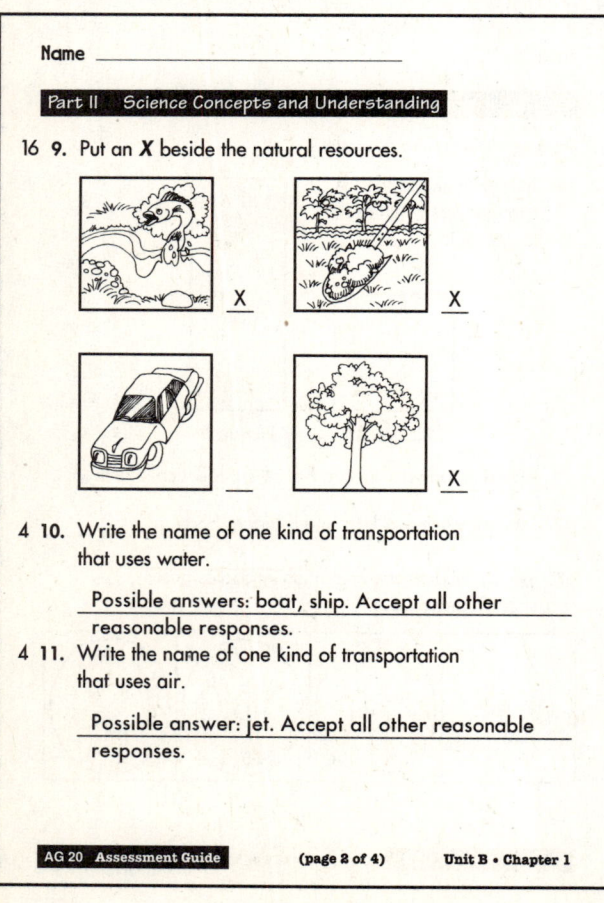

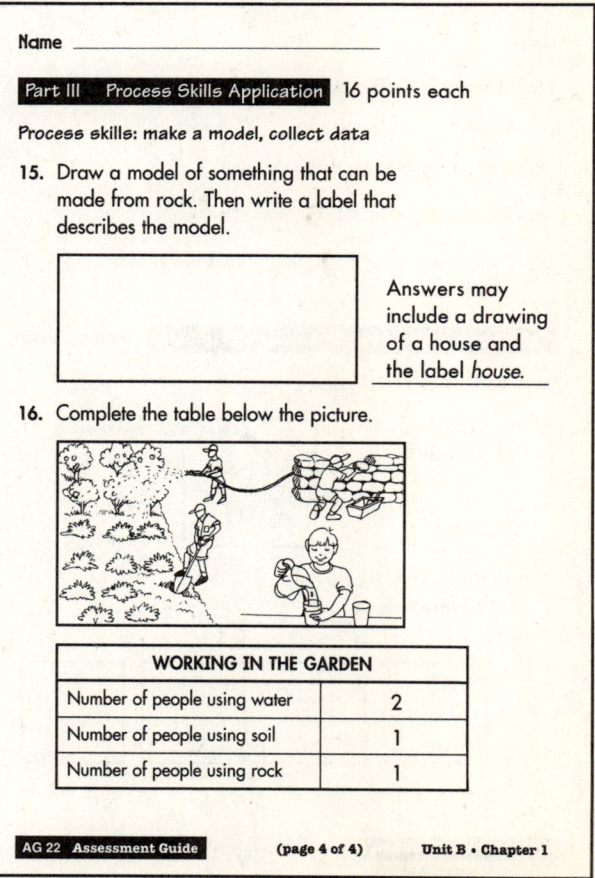

AG 70 Assessment Guide **Answer Key**

Unit B • Chapter 2 • Earth Long Ago

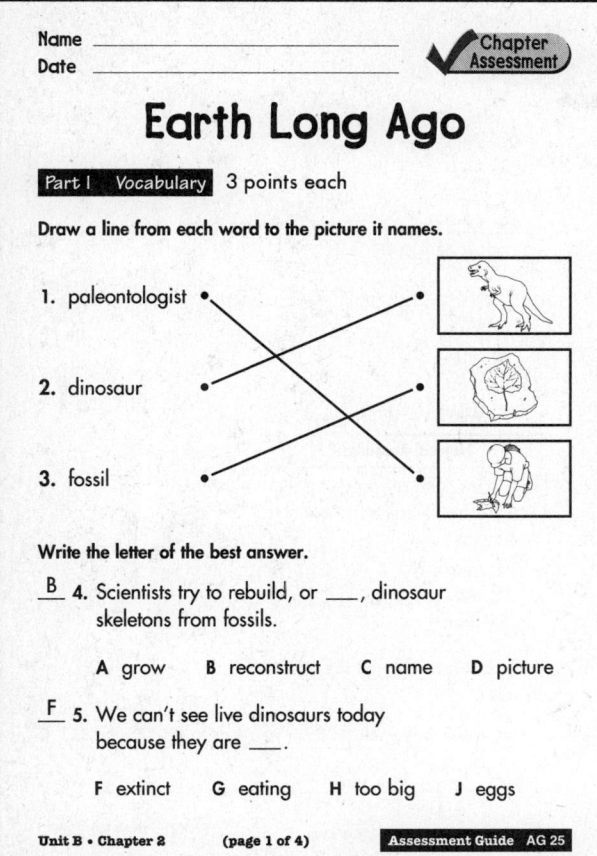

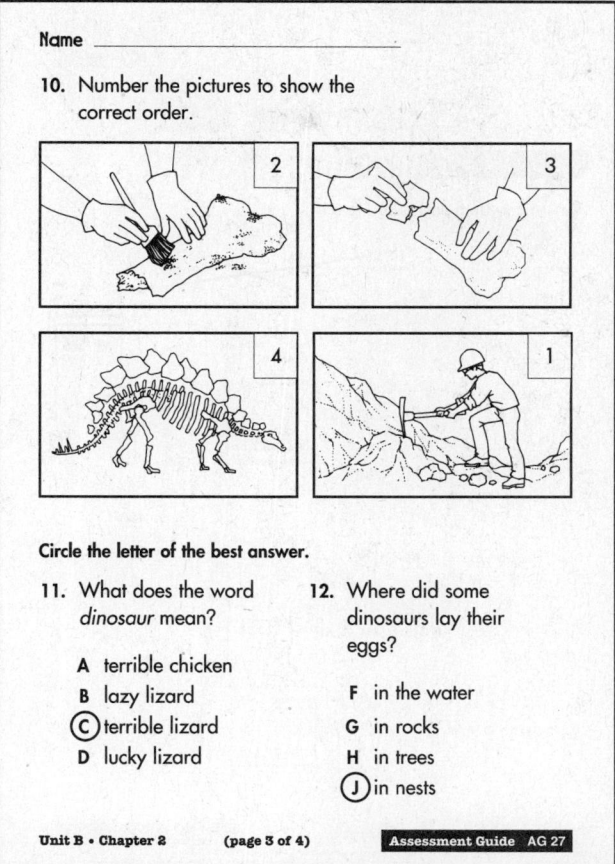

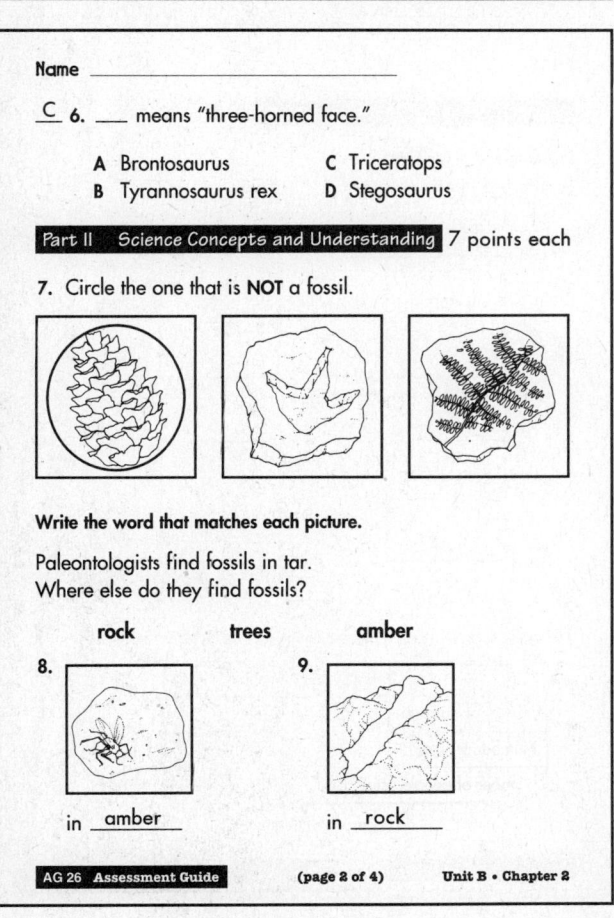

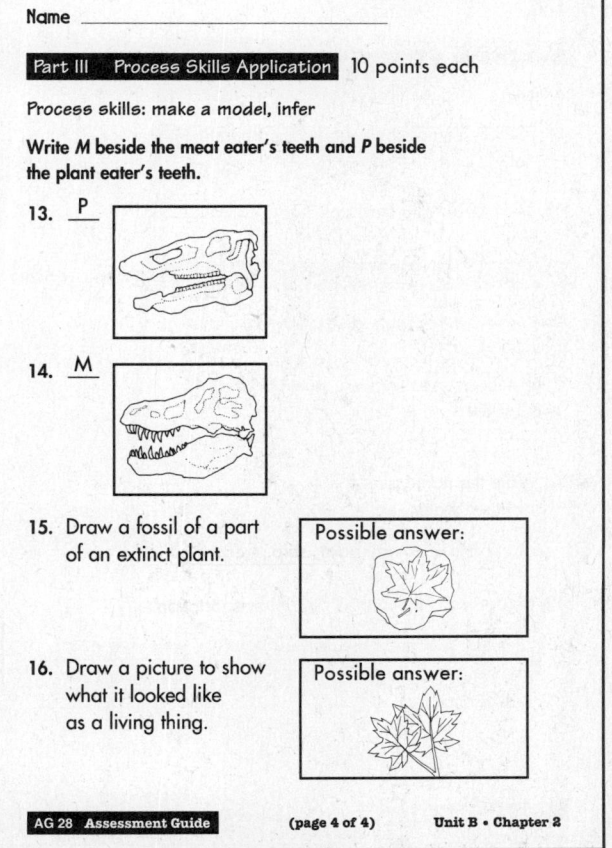

Answer Key

Assessment Guide AG 71

Unit C • Chapter 1 • Observing and Measuring Matter

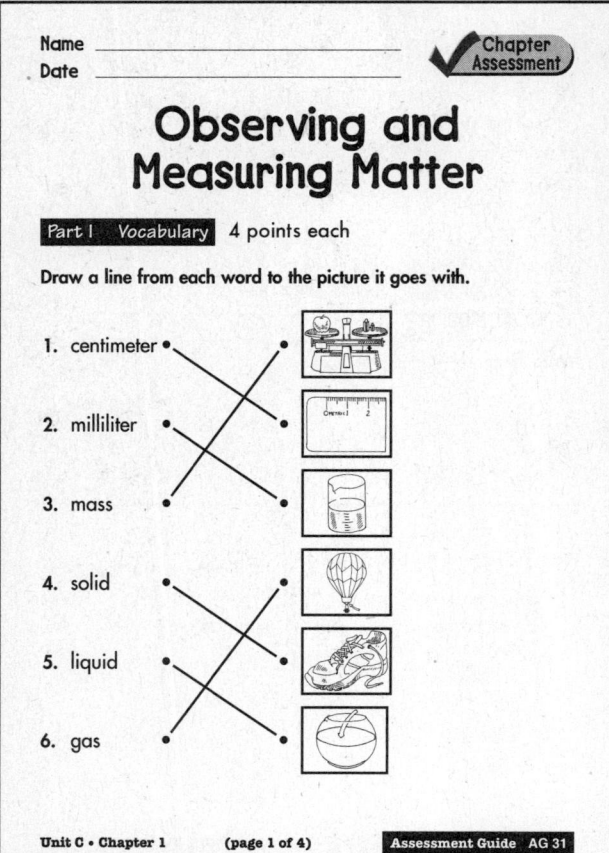

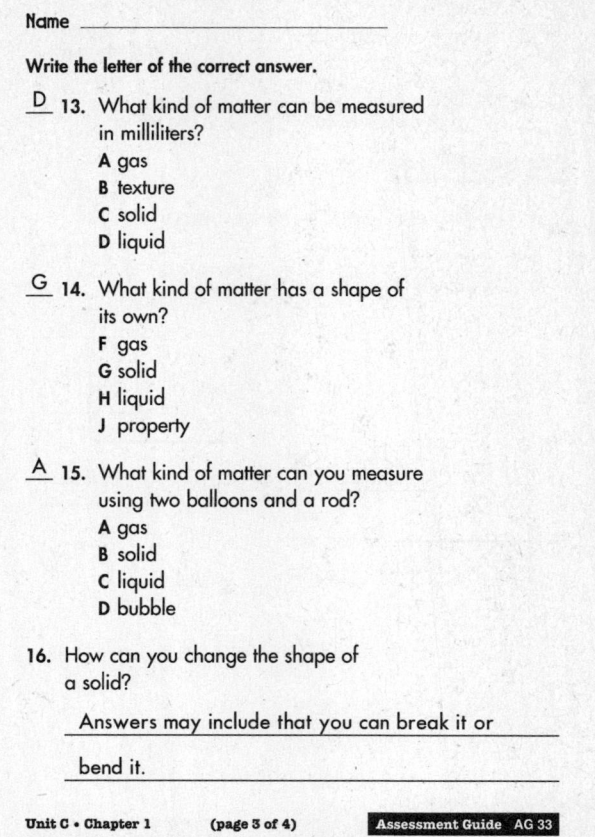

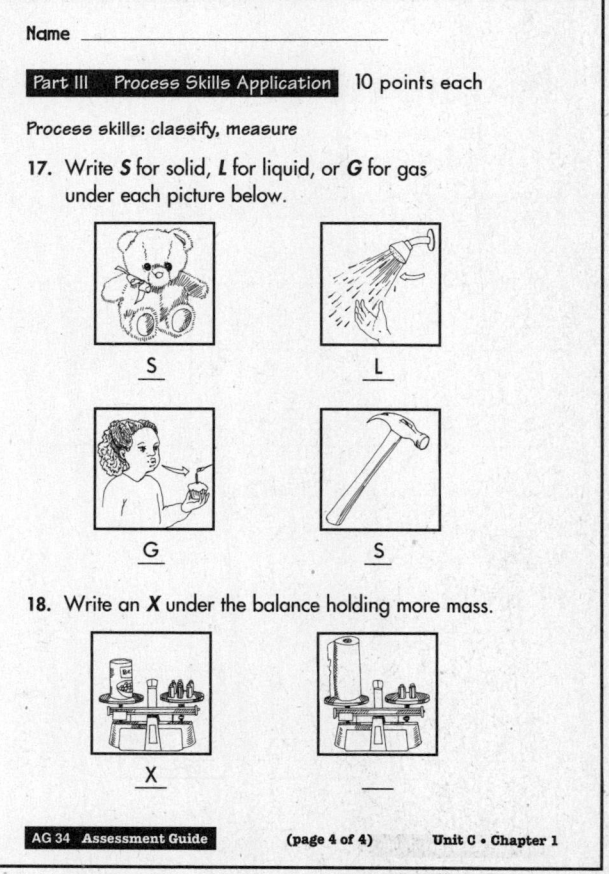

AG 72 Assessment Guide

Answer Key

Unit C • Chapter 2 • Changes in Matter

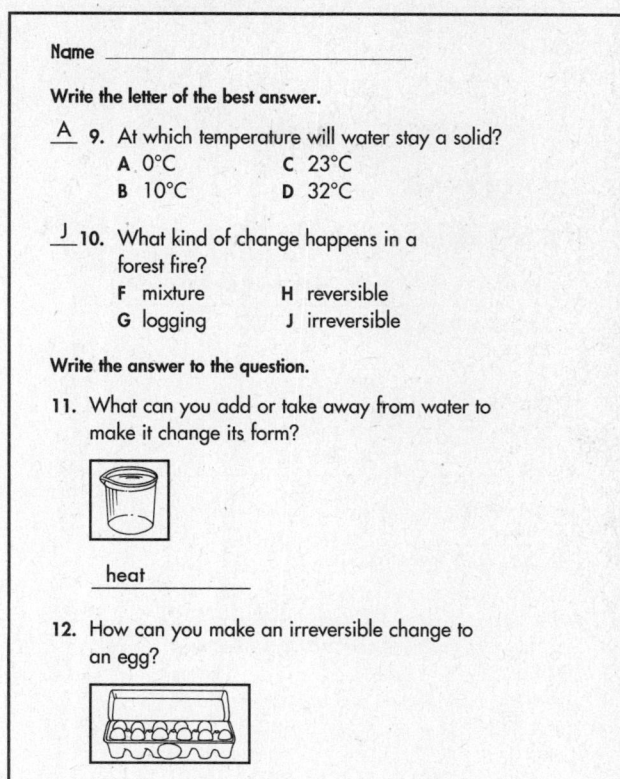

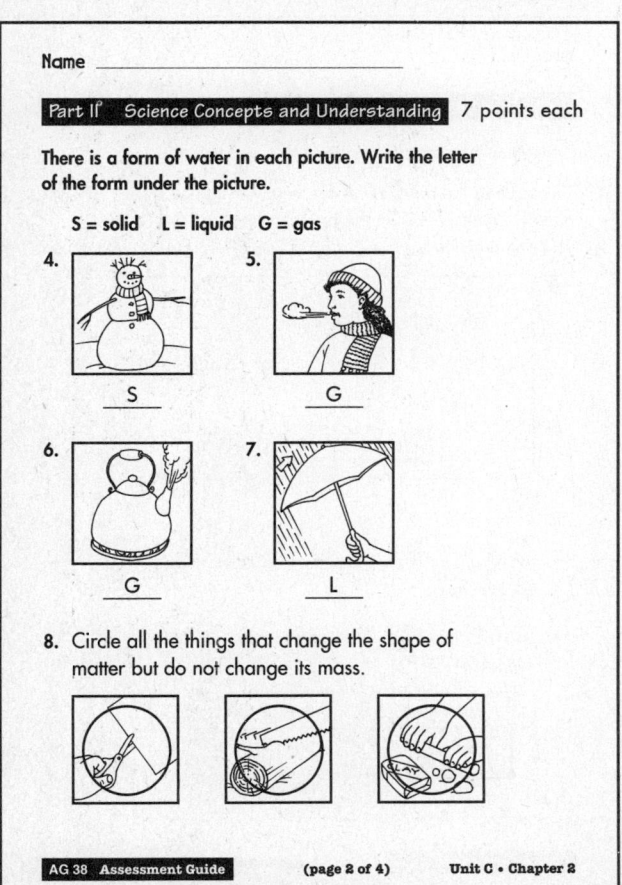

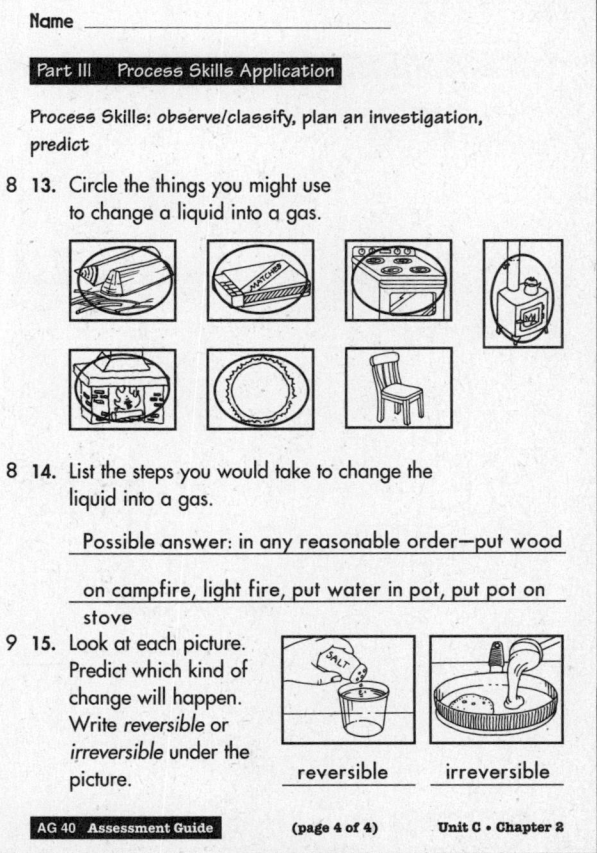

Answer Key

Assessment Guide AG 73

Unit C • Chapter 3 • Forces and Motion

Forces and Motion

Part I Vocabulary 4 points each

Write the correct word to complete the sentence.

| force | gravity | location | motion | wind |

The girl with the wagon is pulling it, or using 1. **force** to move it. She is changing its 2. **location** by moving it toward the trees. The force of moving air, or 3. **wind**, keeps the kite up in the sky. The force that pulls one end of the seesaw down is called 4. **gravity**. When the seesaw is moving up and down, we say it is in 5. **motion**.

---

9. Write an *X* on the picture that shows something a magnet cannot do.

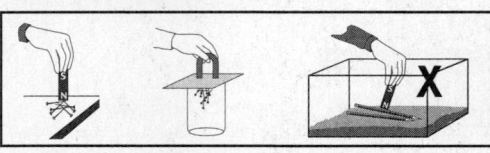

Write the letter of the correct answer.

B 10. What makes it harder to pull a wagon on grass than on a sidewalk?
  A magnetism C wind
  B friction D motion

H 11. What can you use to measure how long it takes to run a race?
  F rubber band H stopwatch
  G meterstick J location

B 12. When you push or pull something, what might you change?
  A its rough surface C its gravity
  B its location D its color

H 13. What is the force that keeps us on the ground?
  F location H gravity
  G wind J grass

Part II Science Concepts and Understanding 7 points each

Write the word that describes the force used in each picture.

6. **wind**

7. **magnetism**

8. **gravity**

Part III Process Skills Application 12 points each

Process skills: observe, use numbers

The scale in the pictures below measures the amount of force. When the hammer hits the scale, you can read the amount of force.

A  B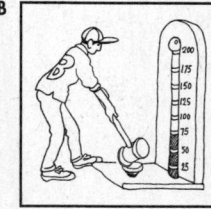

   X   ___

14. Write an *X* below the picture that shows the greater force.

15. Use information from the pictures to complete the table.

Person	Amount of Force
A	200
B	75

Answer Key

Unit C • Chapter 4 • Sound

Sound

Part I Vocabulary 2 points each

Draw a line from the word to the picture it goes with.

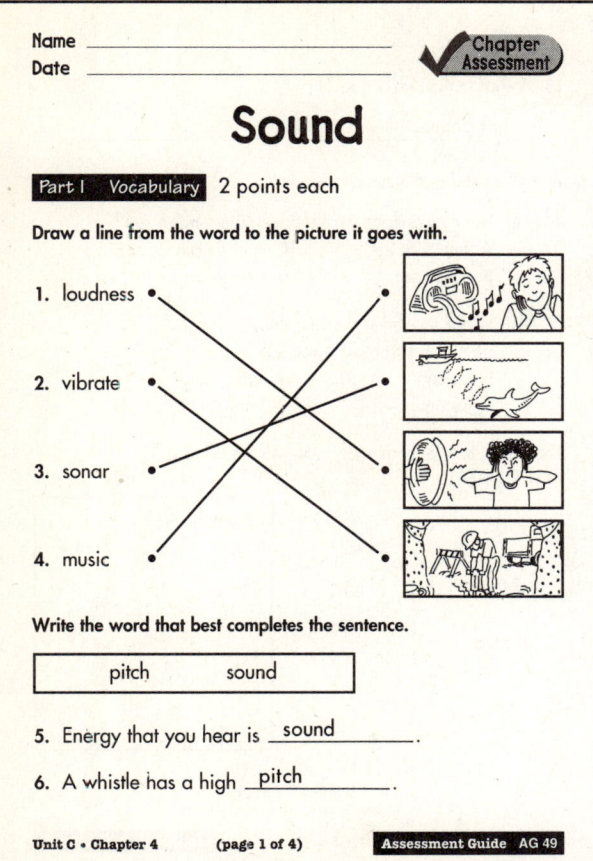

1. loudness
2. vibrate
3. sonar
4. music

Write the word that best completes the sentence.

| pitch | sound |

5. Energy that you hear is ___sound___.
6. A whistle has a high ___pitch___.

Unit C • Chapter 4   (page 1 of 4)   Assessment Guide AG 49

Part II Science Concepts and Understanding 6 points each

7. Why are the plates vibrating?
   The tool is making the air vibrate.

8. Why are the worker's ears covered?
   to protect them from the loud sound

Write the letter of the best choice.

__A__ 9. What vibrates when we speak?
   A vocal cords
   B eardrums
   C inner ears
   D throat

__C__ 11. What vibrates when we hear?
   A throat
   B vocal cords
   C eardrums
   D ear lobes

__G__ 10. What is the pitch of a mouse's squeak?
   F low
   G high
   H soft
   J loud

__G__ 12. Which animal does NOT sing?
   F canary
   G butterfly
   H humpback whale
   J human being

AG 50 Assessment Guide   (page 2 of 4)   Unit C • Chapter 4

Is the sound traveling through a liquid, a solid, or a gas?

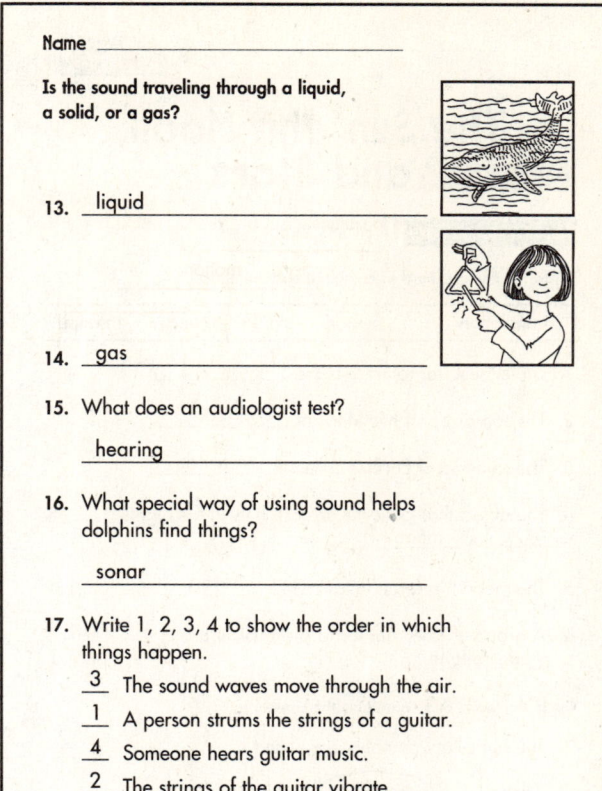

13. ___liquid___

14. ___gas___

15. What does an audiologist test?
    hearing

16. What special way of using sound helps dolphins find things?
    sonar

17. Write 1, 2, 3, 4 to show the order in which things happen.
    __3__ The sound waves move through the air.
    __1__ A person strums the strings of a guitar.
    __4__ Someone hears guitar music.
    __2__ The strings of the guitar vibrate.

Unit C • Chapter 4   (page 3 of 4)   Assessment Guide AG 51

Part III Process Skills Application 11 points each

Process skills: compare, observe, plan an investigation

18. You can tap a solid surface with two parts of a marker or a pencil and compare the sounds. Complete the chart. The two rows for the marker have been done.

SOUND ON A SOLID SURFACE

Tap	Part	Louder	Softer	Higher	Lower
marker	cap		X	X	
	bottom	X			X
pencil	eraser		X		X
	point	X		X	

19. How could you find out whether a large drum or a small drum has a lower pitch? Plan an investigation to answer this question.
    Possible answer: Cover the ends of two sizes of cans with paper. Hold the paper in place with rubber bands. Then tap each drum and listen to the sound.

AG 52 Assessment Guide   (page 4 of 4)   Unit C • Chapter 4

Answer Key   **Assessment Guide AG 75**

Extension • Chapter 1 • The Sun, the Moon, and Stars

The Sun, the Moon, and Stars

Part I Vocabulary   4 points each

Write the correct word to complete each sentence.

| constellation | sun | moon | craters | energy | rotation |

1. The closest star to Earth is the __sun__.
2. The sun gives off two kinds of __energy__.
3. The spinning of Earth is called its __rotation__.
4. The object that looks the largest to us in the night sky is the __moon__.
5. The moon has large holes called __craters__.
6. A group of stars that forms a star picture is a __constellation__.

Circle the word that completes the sentence.

7. The sun's light and heat are called ___.

   orbit   (solar energy)   craters

---

13. What is the sun made of?
    __hot gases__

Write the letter of the correct answer.

__D__ 14. How long does it take Earth to orbit the sun?
   A ten minutes   C one month
   B one hour   D one year

__G__ 15. What season is it when the sun hits our part of Earth most directly?
   F spring   H fall
   G summer   J winter

16. Draw a line from each word to the thing it describes.

   rocks
   astronaut
   crater

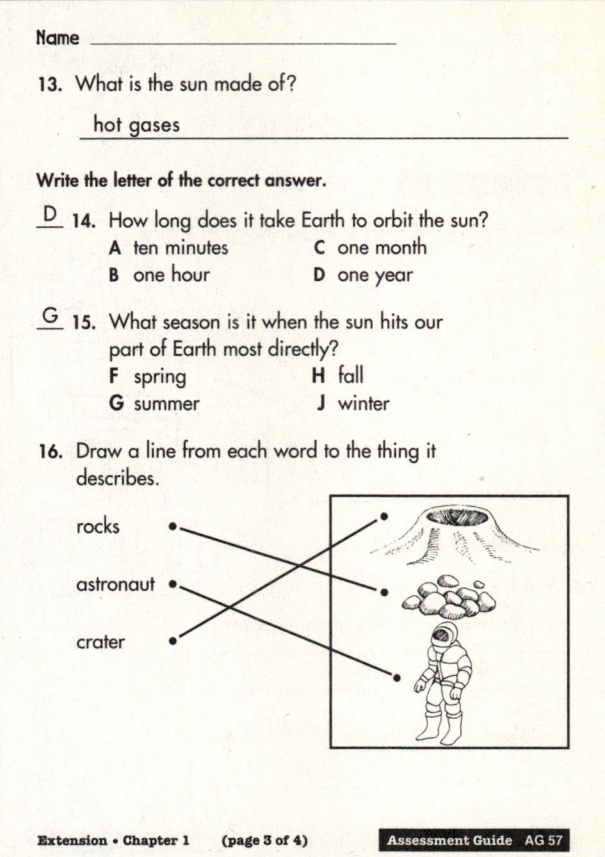

---

8. Earth moves in a path, or an ___, around the sun.

   astronaut   energy   (orbit)

9. Different ___ have different kinds of weather.

   constellations   (seasons)   craters

10. The light from the moon, or ___, comes from the sun.

    (moonlight)   dust   rotation

Part II Science Concepts and Understanding   5 points each

For Questions 11–12, use the picture of Earth and the sun.

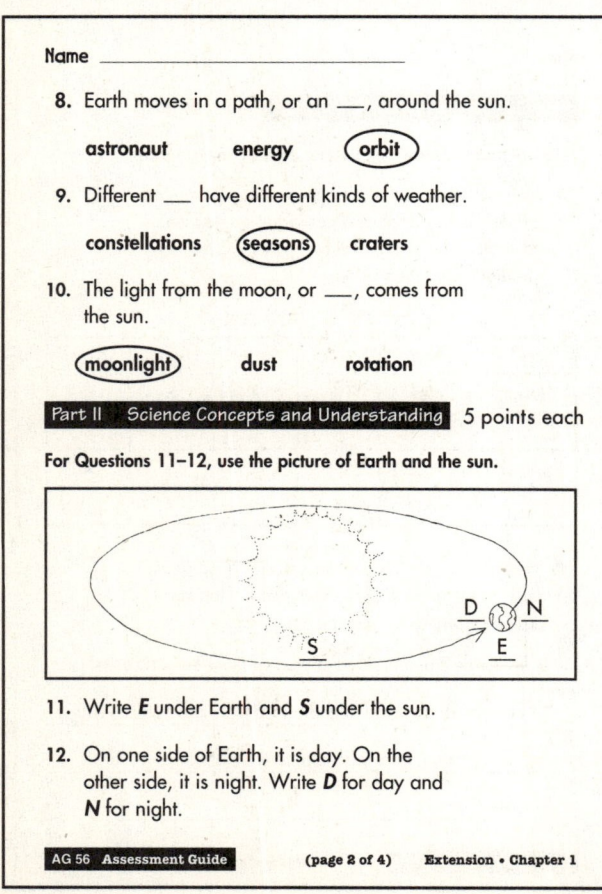

11. Write **E** under Earth and **S** under the sun.
12. On one side of Earth, it is day. On the other side, it is night. Write **D** for day and **N** for night.

Part III Process Skills Application   15 points each

Process skills: time/space relationships, compare

17. Write an **X** under the new moon.

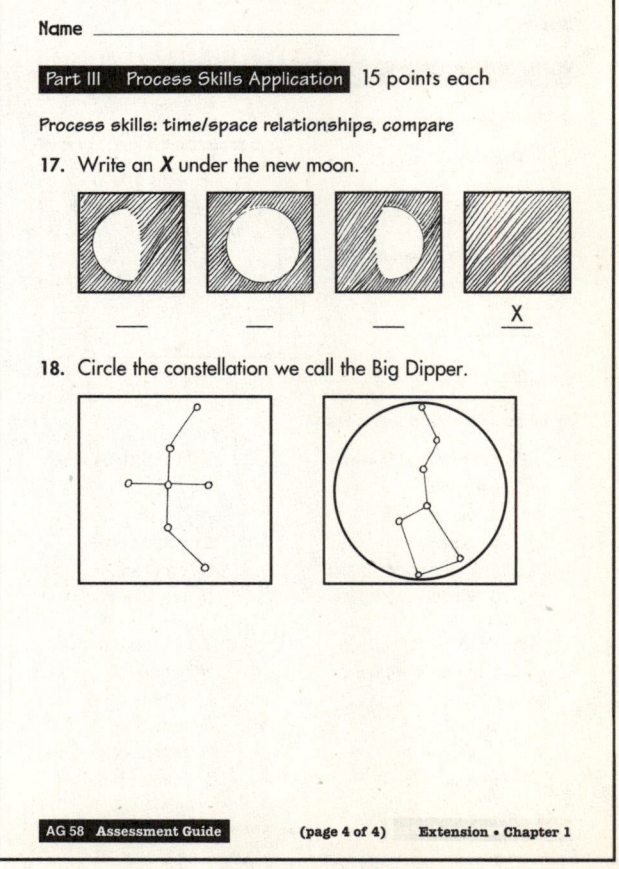
___   ___   ___   _X_

18. Circle the constellation we call the Big Dipper.

Extension • Chapter 2 • Changes in Habitats

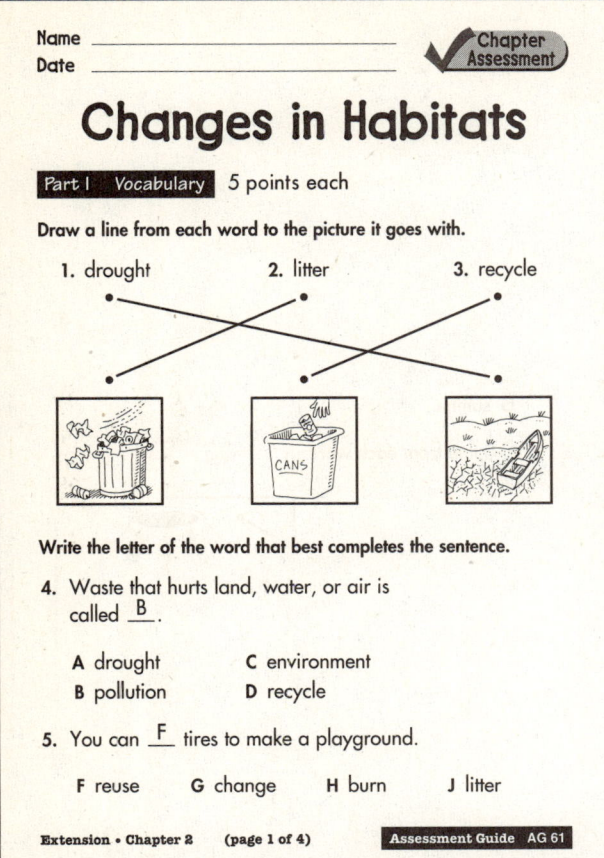

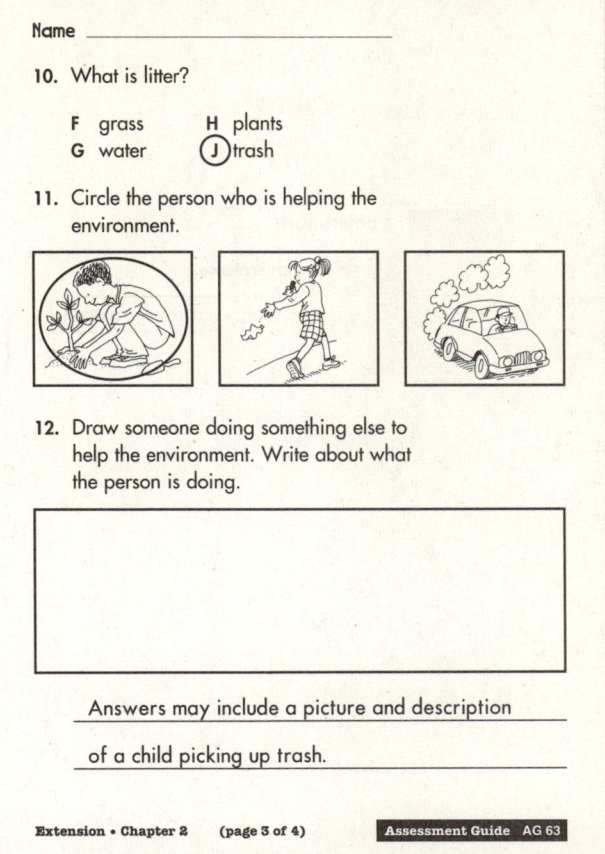

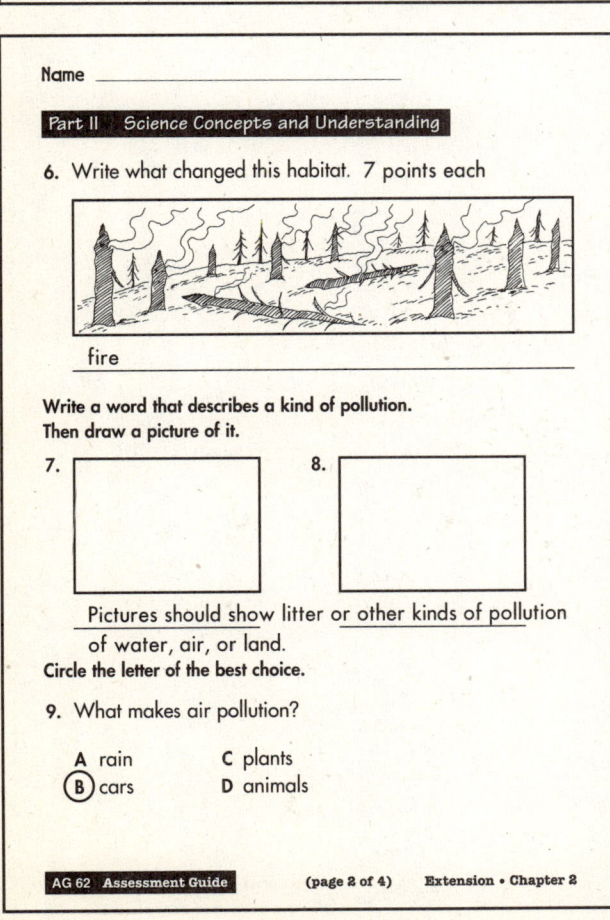

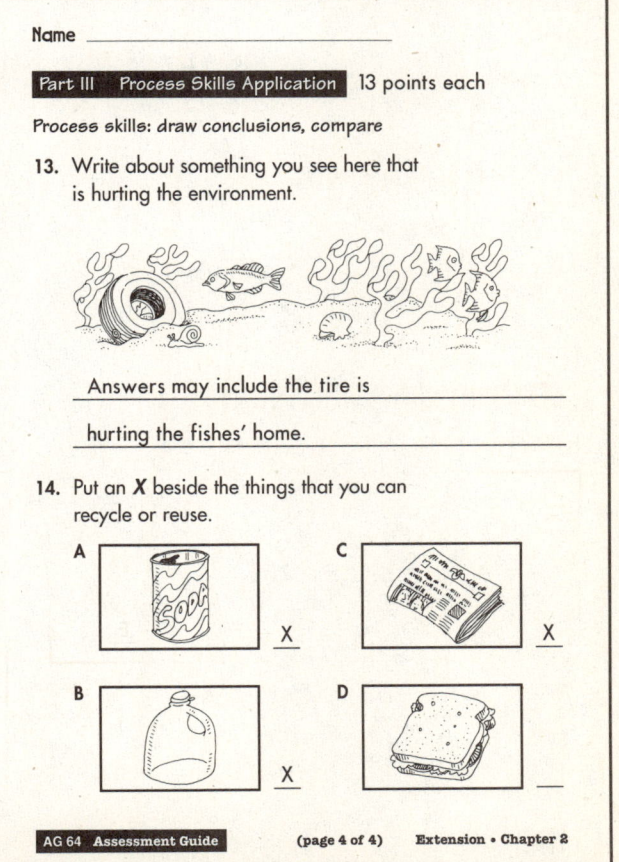

Answer Key

Assessment Guide AG 77